## Microwave and Wireless Measurement Techniques

From typical metrology parameters for common wireless and microwave components to the implementation of measurement benches, this introduction to metrology contains all the key information on the subject. Using it, readers will be able to

- interpret and measure most of the parameters described in a microwave component's datasheet
- understand the practical limitations and theoretical principles of instrument operation
- combine several instruments into measurement benches for measuring microwave and wireless quantities

Several practical examples are included, demonstrating how to measure intermodulation distortion, error-vector magnitude, *S*-parameters, and large-signal waveforms. Each chapter ends with a set of exercises, allowing readers to test their understanding of the material covered and making the book equally suited for course use and for self-study.

**Nuno Borges Carvalho** is a Full Professor at the Universidade de Aveiro, Portugal, and a Senior Research Scientist at the Instituto de Telecomunicações. His main research interests include nonlinear distortion analysis in emerging microwave/wireless circuits and systems, and measurement of nonlinear phenomena.

**Dominique Schreurs** is a Full Professor at the KU Leuven. Her main research interests concern the nonlinear characterization and modeling of microwave devices and circuits, as well as nonlinear hybrid and integrated circuit design for telecommunications and biomedical applications.

The Cambridge RF and Microwave Engineering Series

Series Editor
Steve C. Cripps, Distinguished Research Professor, Cardiff University

Peter Aaen, Jaime Plá, and John Wood, *Modeling and Characterization of RF and Microwave Power FETs*
Dominique Schreurs, Máirtín O'Droma, Anthony A. Goacher, and Michael Gadringer, *RF Amplifier Behavioral Modeling*
Fan Yang and Yahya Rahmat-Samii, *Electromagnetic Band Gap Structures in Antenna Engineering*
Enrico Rubiola, *Phase Noise and Frequency Stability in Oscillators*
Earl McCune, *Practical Digital Wireless Signals*
Stepan Lucyszyn, *Advanced RF MEMS*
Patrick Roblin, *Nonlinear RF Circuits and the Large-Signal Network Analyzer*
Matthias Rudolph, Christian Fager, and David E. Root, *Nonlinear Transistor Model Parameter Extraction Techniques*
John L. B. Walker, *Handbook of RF and Microwave Solid-State Power Amplifiers*
Sorin Voinigescu, *High-Frequency Integrated Circuits*
Valeria Teppati, Andrea Ferrero, and Mohamed Sayed, *Modern RF and Microwave Measurement Techniques*
David E. Root, Jan Verspecht, Jason Horn, and Mihai Marcu, *X-Parameters*
Nuno Borges Carvalho and Dominique Scheurs, *Microwave and Wireless Measurement Techniques*

Forthcoming
Richard Carter, *Theory and Design of Microwave Tubes*
Ali Darwish, Slim Boumaiza, and H. Alfred Hung, *GaN Power Amplifier and Integrated Circuit Design*
Hossein Hashemi and Sanjay Raman, *Silicon mm-Wave Power Amplifiers and Transmitters*
Earl McCune, *Dynamic Power Supply Transmitters*
Isar Mostafanezad, Olga Boric-Lubecke, and Jenshan Lin, *Medical and Biological Microwave Sensors and Systems*

# Microwave and Wireless Measurement Techniques

NUNO BORGES CARVALHO
Universidade de Aveiro, Portugal

DOMINIQUE SCHREURS
KU Leuven, Belgium

CAMBRIDGE
UNIVERSITY PRESS

CAMBRIDGE
UNIVERSITY PRESS

Shaftesbury Road, Cambridge CB2 8EA, United Kingdom

One Liberty Plaza, 20th Floor, New York, NY 10006, USA

477 Williamstown Road, Port Melbourne, VIC 3207, Australia

314–321, 3rd Floor, Plot 3, Splendor Forum, Jasola District Centre, New Delhi – 110025, India

103 Penang Road, #05–06/07, Visioncrest Commercial, Singapore 238467

Cambridge University Press is part of Cambridge University Press & Assessment, a department of the University of Cambridge.

We share the University's mission to contribute to society through the pursuit of education, learning and research at the highest international levels of excellence.

www.cambridge.org
Information on this title: www.cambridge.org/9781107004610



First published 2013

*A catalogue record for this publication is available from the British Library*

*Library of Congress Cataloging-in-Publication data*
Carvalho, Nuno Borges.
Microwave and wireless measurement techniques /
Nuno Borges Carvalho and Dominique Schreurs.
pages cm
Includes bibliographical references and index.
ISBN 978-1-107-00461-0 (Hardback)
1. Microwave measurements. 2. Microwave circuits–Testing. 3. Wireless communication systems–Testing. I. Schreurs, Dominique. II. Title.
TK7876.C397 2013
621.384028′7–dc23 2013013376

ISBN 978-1-107-00461-0 Hardback

# Contents

"This is a text book that every practicing engineer would like to carry into the RF/ microwave laboratory. Written by two known experts in microwave nonlinear measurements, it covers the wide spectrum of microwave instrumentation from the basic definitions of the circuit's figures of merit to the more evolved and up-to-date material of digital/analog and time/frequency instruments and excitation design."
*Jose Carlos Pedro, Universidade de Aveiro, Portugal*

"This book provides an excellent foundation for those wanting to know about contemporary measurement techniques used in wireless and microwave applications. The authors have used both their considerable knowledge of the subject matter along with many years' teaching experience to provide a clear and structured approach to these subject areas. This book is therefore ideally suited as a foundation text for lectures and/or training courses in this area, aimed at graduate level students and professional engineers working in this industry."
*Nick Ridler, IET Fellow*

"Comprehensive, focussed and immediately useful, this book is an excellent resource for all engineers who want to understand and measure the performance of wireless components and systems."
*Uwe Arz, Physikalisch-Technische Bundesanstalt (PTB), Germany*

# Preface

Metrology has been the most important aspect of wireless communication ever since the time of Maxwell and Hertz at its very beginning. In fact, the metrology aspects related to radio communications have for decades been one of the driving forces for progress in radio communication. For instance, radar is nothing more than a very good measurement instrument that can be used in important applications such as target identification. Nevertheless, most wireless systems nowadays depend heavily on metrology and thus measurement, for instance new spectrum management, QoS evaluation and green RF technologies, since they are all supported in high-quality wireless radio components. That is why it is important to understand the figures of merit of the main wireless system components, how to measure them, and how to model them. Moreover, with recent advances in software-defined radio, and in future cognitive radio, measurements in time/frequency/analog/digital domains have become a very important problem to microwave and RF engineers. This book is aimed to give engineers and researchers answers at the beginning of their laboratory adventures in microwave, wireless systems and circuits. It can also be used in connection with a graduate class on measuring wireless systems, or a professor can select parts of the book for a class on wireless systems in the broad sense. The main idea is to have a text that allows the correct identification of the quantities to be measured and their meaning, allows one to understand how to measure those quantities, and allows one to understand the differences between excitation signals, and between instruments, and between quantities to be measured in different domains (time, frequency, analog, and digital). Along this path to completeness the authors expect to give an overview of the main quantities and figures of merit that can be measured, how to measure them, how to calibrate the instruments, and, finally, how to understand the measurement results. Measurements in different domains will also be explained, including the main drawbacks of each approach. The book will thus be organized as follows.

In Chapter 1 the idea is to present to the reader the main important instrumentation and measurable figures of merit that are important for wireless transceivers. We hope that the reader will understand precisely these main figures of merit and the strategies for characterization and modeling. Some information that will facilitate the reading of typical commercial datasheets and the understanding of the most important figures of merit presented on those documents will also be given.

In chapter 2 the instrumentation typically used for wireless transceiver characterization will be presented, especially that involved with radio signals. The instruments will be presented considering the typical figures of merit described in Chapter 1.

No measurement instrumentation can work without the need for appropriate excitation, so in Chapter 3 we will present the most important excitations for radio characterization, namely those mainly supported in sinusoidal excitations.

In Chapter 4 the main idea is to present several test benches for modeling and characterization that allow a correct identification of several linear and nonlinear parameters useful for wireless systems.

The work of writing and publishing this book is not exclusively that of the authors, but includes the help and collaboration of many persons who came into our lives during this process of duration almost 4 years. So we would like to express our gratitude to the many people who, directly or indirectly, helped us to carry out this task.

The first acknowledgments go to our families for their patience and emotional support. In addition we are especially in debt to a group of our students, or simply collaborators, who contributed some results, images, and experimental data to the book. They include Pedro Miguel Cruz, Diogo Ribeiro, Paulo Gonçalves, Hugo Cravo Gomes, Alirio Boaventura, Hugo Mostardinha, Maciej Myśliński, and Gustavo Avolio, among others. Finally we would like to acknowledge Dr. Kate Remley from the National Institute of Standards and Technology (NIST) for the initial ideas for the book, the financial and institutional support provided by the Portuguese National Science Foundation (FCT), the Instituto de Telecomunicações, Departamento de Electrónica, Telecomunicações e Informática of the Universidade de Aveiro, as well as the FWO-Flanders.

# Notation

| | |
|---|---|
| $(I)$ | unit matrix |
| $\delta(\,)$ | Dirac function |
| $\eta$ | efficiency |
| $\eta_e$ | effective efficiency |
| $\Gamma(x)$ | reflection coefficient |
| $\Gamma_{IN}$ | input reflection coefficient |
| $\Gamma_{OPT}$ | optimum source-reflection coefficient for minimum noise figure |
| $\Gamma_{OUT}$ | output reflection coefficient |
| $\omega$ | pulsation |
| $\phi(t)$ | modulated phase |
| $\sigma$ | error variance |
| $a$ | incident traveling voltage wave |
| ACPR | adjacent-channel power ratio |
| $ACPR_L$ | lower adjacent-channel power ratio |
| $ACPR_T$ | total adjacent-channel power ratio |
| $ACPR_U$ | upper adjacent-channel power ratio |
| $ACPR_{SP}$ | spot adjacent-channel power ratio |
| $b$ | scattered traveling voltage wave |
| $C_S$ | correlation matrix in terms of $S$-parameters |
| $C_Y$ | correlation matrix in terms of $Y$-parameters |
| DR | dynamic range |
| $E(t)$ | energy variation over time |
| $E_s$ | energy source |
| $F$ | noise factor |
| $f$ | frequency |
| $f_0$ | fundamental frequency |
| $F_{MIN}$ | minimum noise factor |
| $G$ | operating power gain |
| $G_A$ | available power gain |
| $G_T$ | transducer power gain |
| $I$ | current |
| $i$ | instantaneous current |
| $i(t)$ | instantaneous current over time |
| $I_D$ | diode current |

| | |
|---|---|
| $I_{D0}$ | diode bias current |
| $IIP_3$ | input third-order intercept point |
| $IP_3$ | third-order intercept point |
| $IP_{1dB}$ | 1-dB-compression point referred to the input |
| $k_B$ | Boltzmann constant |
| $L$ | conversion loss |
| NF | noise figure |
| $P(t)$ | power variation over time |
| $P_R$ | power dissipated over a resistance $R$ |
| $P_s$ | source power |
| $P_{1dB}$ | 1-dB-compression point |
| $P_{DC}$ | DC power |
| $P_{diss}$ | dissipated power |
| $P_{gi}$ | incident measured power |
| $P_{IMD}$ | intermodulation distortion power |
| $P_i$ | incident power |
| $P_{LA}$ | total power in lower adjacent-channel band |
| $P_{sat}$ | saturated output power |
| $P_{UA}$ | total power in upper adjacent-channel band |
| $q$ | charge of an electron |
| $R$ | resistance |
| $R_N$ | noise resistance |
| $S$-parameters | scattering parameters |
| $S_i$ | sensitivity |
| ST | sweep time |
| $T$ | temperature |
| $T$ | time window (period) |
| $t_r$ | rise time |
| $V$ | voltage |
| $v$ | instantaneous voltage |
| $v(t)$ | instantaneous voltage over time |
| $V_D$ | diode voltage |
| $V_T$ | thermal voltage |
| $V_{D_{Bias}}$ | diode bias voltage |
| $Y$-parameters | admittance parameters |
| $Y_{OPT}$ | optimum source admittance for minimum noise figure |
| $Z$-parameters | impedance parameters |
| $Z_0$ | characteristic impedance |
| $Z_L$ | load impedance |
| $Z_S$ | source impedance |

# Abbreviations

| | |
|---|---|
| ADC | analog-to-digital converter |
| AM | amplitude modulation |
| ASIC | application-specific integrated circuit |
| AWG | arbitrary-waveform generator |
| BPF | bandpass filter |
| ccdf | complementary cumulative distribution function |
| CCPR | co-channel power ratio |
| CF | calibration factor |
| CR | cognitive radio |
| CRT | cathode-ray tubes |
| CW | carrier wave |
| DAC | digital-to-analog converter |
| dB | decibel |
| dBc | decibels below carrier |
| DC | direct current |
| DDS | direct digital synthesis |
| DFS | direct frequency synthesis |
| DSP | digital signal processor |
| DUT | device under test |
| emf | electromotive force |
| ENBW | equivalent noise bandwidth |
| ENOB | effective number of bits |
| EVM | error-vector magnitude |
| FDM | frequency-division multiplex |
| FFT | fast fourier transform |
| FOM | figure of merit |
| FPGA | field-programmable gate array |
| GA | generic amplifier |
| IF | intermediate frequency |
| IIP3 | third-order intercept point referred to the input |
| IMD | intermodulation distortion |
| IMR | intermodulation ratio |
| LNA | low-noise amplifier |
| LO | local oscillator |

| | |
|---|---|
| LSB | least significant bit |
| M-IMR | multi-sine intermodulation ratio |
| MIMO | multiple input, multiple output |
| MSB | most-significant bit |
| NBGN | narrowband Gaussian noise |
| NPR | noise power ratio |
| NVNA | nonlinear vector network analyzer |
| OCXO | oven-controlled oscillator |
| OFDM | orthogonal frequency-division multiplexing |
| OSR | over-sample ratio |
| PA | power amplifier |
| PAE | power added efficiency |
| PAPR | peak-to-average power ratio |
| pdf | probability density function |
| PLL | phase-locked loop |
| PM | phase modulation |
| PSD | power spectral density |
| QPSK | quadrature phase-shift keying |
| RBW | resolution bandwidth |
| RF | radio frequency |
| RL | return loss |
| RMS | root mean square |
| RTSA | real-time signal analyzer |
| SA | spectrum analyzer |
| SDR | software-defined radio |
| SFDR | spurious free dynamic range |
| SINAD | signal-to-noise-and-distortion |
| SNR | signal-to-noise ratio |
| SNR_ADC | signal-to-noise ratio for ADCs |
| SSB noise | single-sideband noise |
| STFT | short-time Fourier transform |
| SUT | system under test |
| TCXO | temperature-compensated crystal oscillator |
| TDD | time-division duplex |
| THD | total harmonic distortion |
| ULG | underlying linear gain |
| VCO | voltage-controlled oscillator |
| VGA | variable-gain amplifier |
| VNA | vector network analyzer |
| VSA | vector signal analyzer |
| VSWR | voltage standing-wave ratio |
| YIG | yttrium iron garnet |

# 1 Measurement of wireless transceivers

## 1.1 Introduction

This book is entitled *Microwave and Wireless Measurement Techniques*, since the objective is to identify and understand measurement theory and practice in wireless systems.

In this book, the concept of a wireless system is applied to the collection of sub-systems that are designed to behave in a particular way and to apply a certain procedure to the signal itself, in order to convert a low-frequency information signal, usually called the baseband signal, to a radio-frequency (RF) signal, and transmit it over the air, and vice versa.

Figure 1.1 presents a typical commercial wireless system architecture. The main blocks are amplifiers, filters, mixers, oscillators, passive components, and domain converters, namely digital to analog and vice versa.

In each of these sub-systems the measurement instruments will be measuring voltages and currents as in any other electrical circuit. In basic terms, what we are measuring are always voltages, like a voltmeter will do for low-frequency signals. The problem here is stated as how we are going to be able to capture a high-frequency signal and identify and quantify its amplitude or phase difference with a reference signal. This is actually the problem throughout the book, and we will start by identifying the main figures of merit that deserve to be measured in each of the identified sub-systems.

In order to do that, we will start by analyzing a general sub-system that can be described by a network. In RF systems it can be a single-port, two-port, or three-port network. The two-port network is the most common.

## 1.2 Linear two-port networks

### 1.2.1 Microwave description

A two-port network, Fig. 1.2, is a network in which the terminal voltages and currents relate to each other in a certain way.

The relationships between the voltages and currents of a two-port network can be given by matrix parameters such as $Z$-parameters, $Y$-parameters, or ABCD parameters. The reader can find more information in [1, 2].

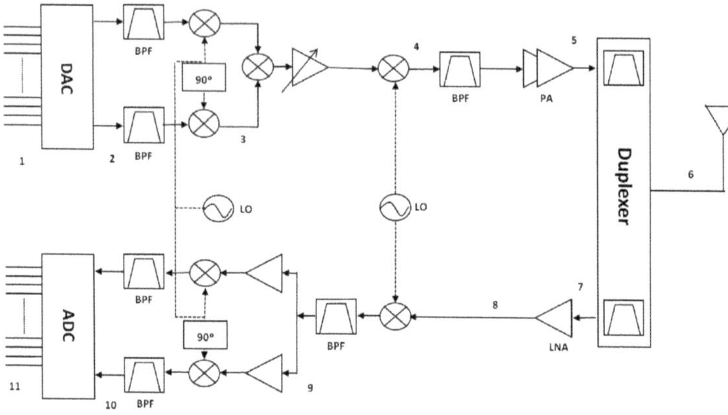


**Figure 1.1** A typical wireless system architecture, with a full receiver and transmitter stage.

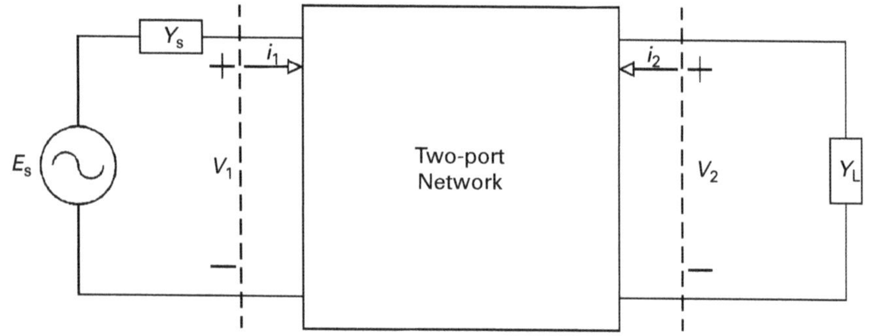


**Figure 1.2** A two-port network, presenting the interactions of voltages and currents at its ports.

The objective is always to relate the input and output voltages and currents by using certain relationships. One of these examples using $Y$-parameters is described by the following equation:

$$\begin{bmatrix} i_1 \\ i_2 \end{bmatrix} = \begin{bmatrix} y_{11} & y_{12} \\ y_{21} & y_{22} \end{bmatrix} \begin{bmatrix} v_1 \\ v_2 \end{bmatrix} \tag{1.1}$$

where

$$y_{11} = \left. \frac{i_1}{v_1} \right|_{v_2=0}$$

$$y_{12} = \left. \frac{i_1}{v_2} \right|_{v_1=0}$$

$$y_{21} = \left. \frac{i_2}{v_1} \right|_{v_2=0}$$

$$y_{22} = \left. \frac{i_2}{v_2} \right|_{v_1=0}$$

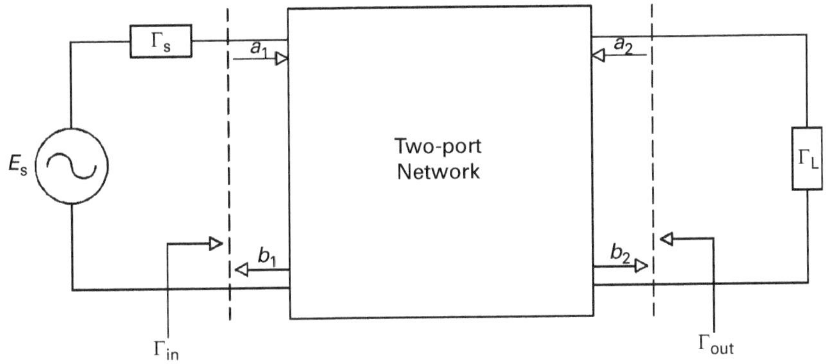


**Figure 1.3** Two-port scattering parameters, where the incident and reflected waves can be seen in each port.

As can be seen, these $Y$-parameters can be easily calculated by considering the other port voltage equal to zero, which means that the other port should be short-circuited. For instance, $y_{11}$ is the ratio of the measured current at port 1 and the applied voltage at port 1 by which port 2 is short-circuited.

Unfortunately, when we are dealing with high-frequency signals, a short circuit is not so simple to realize, and in that case more robust high-frequency parameters should be used.

In that sense some scientists started to think of alternative ways to describe a two-port network, and came up with the idea of using traveling voltage waves [1, 2]. In this case there is an incident traveling voltage wave and a scattered traveling voltage wave at each port, and the network parameters become a description of these traveling voltage waves, Fig. 1.3.

One of the most well-known matrices used to describe these relations consists of the scattering parameters, or $S$-parameters, by which the scattered traveling voltage waves are related to the incident traveling voltage waves in each port.

In this case each voltage and current in each port will be divided into an incident and a scattered traveling voltage wave, $V^+(x)$ and $V^-(x)$, where the $+$ sign refers to the incident traveling voltage wave and the $-$ sign refers to the reflected traveling voltage wave. The same can be said about the currents, where $I^+(x) = V^+(x)/Z_0$ and $I^-(x) = V^-(x)/Z_0$, $Z_0$ being the characteristic impedance of the port. The value $x$ now appears since we are dealing with waves that travel across the space, being guided or not, so $V^+(x) = Ae^{-\gamma x}$ [1, 2].

These equations can be further simplified and normalized to be used efficiently:

$$\begin{aligned} v(x) &= \frac{V(x)}{\sqrt{Z_0}} \\ i(x) &= \sqrt{Z_0} I(x) \end{aligned} \tag{1.2}$$

Then each normalized voltage and current can be decomposed into its incident and scattered wave. The incident wave is denoted $a(x)$ and the scattered one $b(x)$:

$$v(x) = a(x) + b(x)$$
$$i(x) = a(x) - b(x) \tag{1.3}$$

where

$$a(x) = \frac{V^+(x)}{\sqrt{Z_0}}$$
$$b(x) = \frac{V^-(x)}{\sqrt{Z_0}} \tag{1.4}$$

with

$$V = \sqrt{Z_0}(a+b)$$
$$I = \frac{1}{\sqrt{Z_0}}(a-b)$$

Fortunately, we also know that in a load the reflected wave can be related to the incident wave using its reflection coefficient $\Gamma(x)$:

$$b(x) = \Gamma(x)a(x)$$

or

$$\Gamma(x) = \frac{b(x)}{a(x)} \tag{1.5}$$

In this way it is then possible to calculate and use a new form of matrix parameter to describe these wave relationships in a two-port network, namely the scattering parameters:

$$\begin{bmatrix} b_1 \\ b_2 \end{bmatrix} = \begin{bmatrix} S_{11} & S_{12} \\ S_{21} & S_{22} \end{bmatrix} \begin{bmatrix} a_1 \\ a_2 \end{bmatrix} \tag{1.6}$$

where

$$S_{ij} = \left. \frac{b_i(x)}{a_j(x)} \right|_{a_k=0 \text{ to } k \neq j} \tag{1.7}$$

As can be deduced from the equations, and in contrast to the $Y$-parameters, for the calculation of each parameter, the other port should have no reflected wave. This corresponds to matching the other port to the impedance of $Z_0$. This is easier to achieve at high frequencies than realizing a short circuit or an open circuit, as used for $Y$- and $Z$-parameters, respectively.

Moreover, using this type of parameter allows us to immediately calculate a number of important parameters for the wireless sub-system. On looking at the next set of equations, it is possible to identify the input reflection coefficient immediately from $S_{11}$, or, similarly, the output reflection coefficient from $S_{22}$:

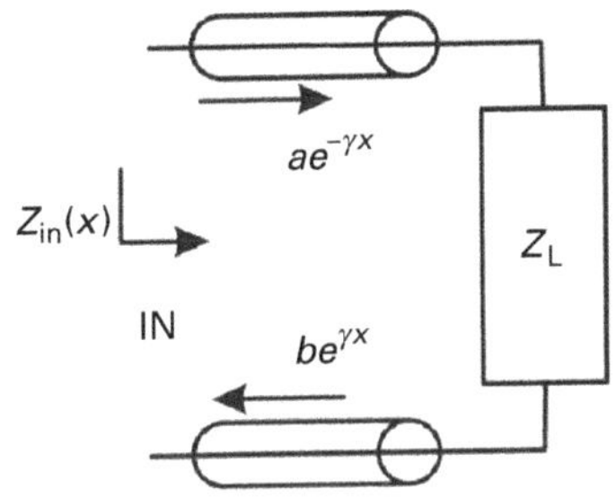


**Figure 1.4** Power waves traversing a guided structure.

$$S_{11} = \left.\frac{b_1(x)}{a_1(x)}\right|_{a_2=0} = \frac{Z_1 - Z_0}{Z_1 + Z_0} \tag{1.8}$$

$$S_{22} = \left.\frac{b_2(x)}{a_2(x)}\right|_{a_1=0} = \frac{Z_2 - Z_0}{Z_2 + Z_0} \tag{1.9}$$

The same applies to the other two parameters, $S_{21}$ and $S_{12}$, which correspond to the transmission coefficient and the reverse transmission coefficient, respectively. The square of their amplitude corresponds to the forward and reverse power gain when the other port is matched.

Note that in the derivation of these parameters it is assumed that the other port is matched. If that is not the case, the values can be somewhat erroneous. For instance, $\Gamma_{in}(x) = S_{11}$ only if the other port is matched or either $S_{12}$ or $S_{21}$ is equal to zero. If this is not the case, the input reflection should be calculated from

$$\Gamma_{in} = S_{11} + \frac{S_{12}S_{21}\Gamma_L}{1 + S_{22}\Gamma_L} \tag{1.10}$$

More information can be found in [1, 2].

With the parameters based on the wave representation that have now been defined, several quantities can be calculated. See Fig. 1.4.

For example, if the objective is to calculate the power at terminal IN, then

$$P = VI^* = aa^* - bb^* = |a|^2 - |b|^2 \tag{1.11}$$

Here $|a|^2$ actually corresponds to the incident power, while $|b|^2$ corresponds to the reflected power.

Important linear figures of merit that are common to most wireless sub-systems can now be defined using the $S$-parameters.

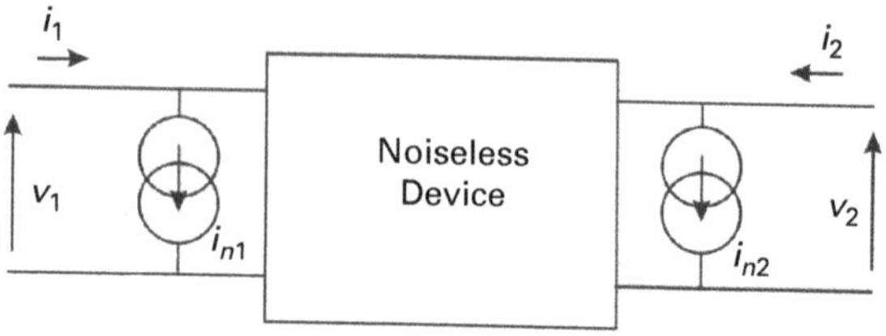


**Figure 1.5** A noisy device, $Y$-parameter representation, including noise sources.

### 1.2.2 Noise

Another very important aspect to consider when dealing with RF and wireless systems is the amount of introduced noise. Since for RF systems the main goal is actually to achieve a good compromise between power and noise, in order to achieve a good noise-to-power ratio, the study of noise is fundamental. For that reason, let us briefly describe the noise behavior [3] in a two-port network.

A noisy two-port network can be represented by a noiseless two-port network and a noise current source at each port. An admittance representation can be developed.

The voltages and currents in each port can be related to the admittance matrix:

$$\begin{bmatrix} i_1 \\ i_2 \end{bmatrix} = [Y] \begin{bmatrix} v_1 \\ v_2 \end{bmatrix} + \begin{bmatrix} i_{n1} \\ i_{n2} \end{bmatrix} \tag{1.12}$$

(Fig. 1.5). A correlation matrix $C_Y$ can also be defined, as

$$[C_Y] = \begin{bmatrix} \langle i_{n1} i_{n1}^* \rangle & \langle i_{n1} i_{n2}^* \rangle \\ \langle i_{n2} i_{n1}^* \rangle & \langle i_{n2} i_{n2}^* \rangle \end{bmatrix} \tag{1.13}$$

The correlation matrix relates the properties of the noise in each port. For a passive two-port network, one has

$$[C_Y] = 4k_B T \, \Delta f \, \mathrm{Re}(Y) \tag{1.14}$$

where $k_B$ is the Boltzmann constant ($1.381 \times 10^{-23}$ J/K), $T$ the temperature (typically 290 K), $\Delta f$ the bandwidth, and $Y$ the admittance parameter.

Actually these port parameters can also be represented by using scattering parameters. In that case the noisy two-port network is represented by a noiseless two-port network and the noise scattering parameters referenced to a nominal impedance at each port (Fig. 1.6).

$$\begin{bmatrix} b_1 \\ b_2 \end{bmatrix} = [S] \begin{bmatrix} a_1 \\ a_2 \end{bmatrix} + \begin{bmatrix} b_{\mathrm{n}1} \\ b_{\mathrm{n}2} \end{bmatrix} \tag{1.15}$$

where $b_{\mathrm{n}1}$ and $b_{\mathrm{n}2}$ can be considered noise waves, and they are related using the correlation matrix, $C_S$. The correlation matrix $C_S$ is defined by

$$[C_S] = \begin{bmatrix} \langle b_{\mathrm{n}1} b_{\mathrm{n}1}^* \rangle & \langle b_{\mathrm{n}1} b_{\mathrm{n}2}^* \rangle \\ \langle b_{\mathrm{n}2} b_{\mathrm{n}1}^* \rangle & \langle b_{\mathrm{n}2} b_{\mathrm{n}2}^* \rangle \end{bmatrix} \tag{1.16}$$

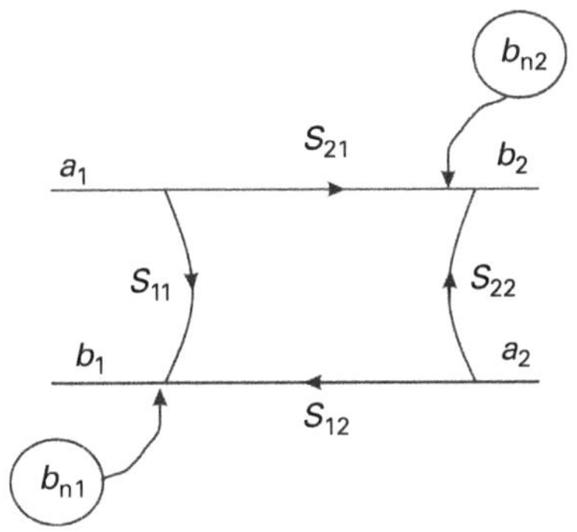


**Figure 1.6** A noisy device, $S$-parameter representation.

and, for a passive two-port network,

$$[C_S] = k_B T \, \Delta f\big((I) - (S)(S)^{T*}\big) \tag{1.17}$$

where $(I)$ is the unit matrix and $(S)^{T*}$ denotes transpose and conjugate.

## 1.3 Linear FOMs

After having described linear networks, we proceed to explain the corresponding figures of merit (FOMs). We make a distinction between FOMs that are defined on the basis of $S$-parameters (Section 1.3.1) and those defined on the basis of noise (Section 1.3.2).

### 1.3.1 Linear network FOMs

#### 1.3.1.1 The voltage standing-wave ratio

The voltage standing-wave ratio (VSWR) is nothing more than the evaluation of the port mismatch. Actually, it is a similar measure of port matching, the ratio of the standing-wave maximum voltage to the standing-wave minimum voltage. Figure 1.7 shows different standing-wave patterns depending on the load.

In this sense it therefore relates the magnitude of the voltage reflection coefficient and hence the magnitude of either $S_{11}$ for the input port or $S_{22}$ for the output port.

The VSWR for the input port is given by

$$\mathrm{VSWR_{in}} = \frac{1 + |S_{11}|}{1 - |S_{11}|} \tag{1.18}$$

and that for the output port is given by

$$\mathrm{VSWR_{out}} = \frac{1 + |S_{22}|}{1 - |S_{22}|} \tag{1.19}$$

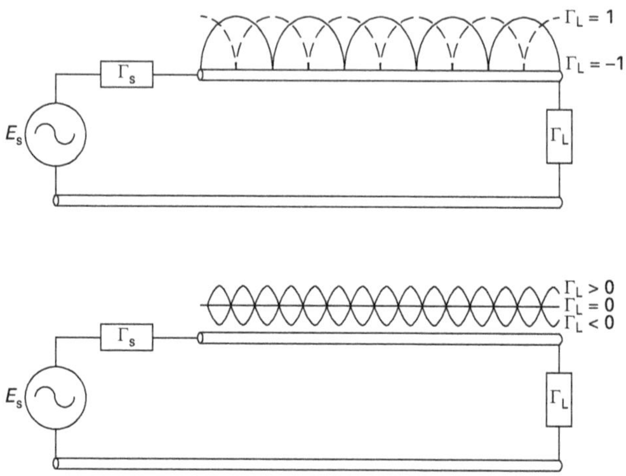


**Figure 1.7** The VSWR and standing-wave representation. The standing wave can be seen for different values of the VSWR.

### 1.3.1.2 Return loss

Other important parameters are the input and output return losses. The input return loss ($RL_{in}$) is a scalar measure of how close the actual input impedance of the network is to the nominal system impedance value, and is given by

$$RL_{in} = \left|20\log_{10}\left|S_{11}\right|\right| \text{dB} \tag{1.20}$$

It should be noticed that this value is valid only for a single-port network, or, in a two-port network, it is valid only if port 2 is matched; if not, $S_{11}$ should be exchanged for the input reflection coefficient as presented in Eq. (1.10). As can be seen from its definition, the return loss is a positive scalar quantity.

The output return loss ($RL_{out}$) is similar to the input return loss, but applied to the output port (port 2). It is given by

$$RL_{out} = \left|20\log_{10}\left|S_{22}\right|\right| \text{dB} \tag{1.21}$$

### 1.3.1.3 Gain/insertion loss

Since $S_{11}$ and $S_{22}$ have the meaning of reflection coefficients, their values are always smaller than or equal to unity. The exception is the $S_{11}$ of oscillators, which is larger than unity, because the RF power returned is larger than the RF power sent into the oscillator port.

The $S_{21}$ of a linear two-port network can have values either smaller or larger than unity. In the case of passive circuits, $S_{21}$ has the meaning of loss, and is thus restricted to values smaller than or equal to unity. This loss is usually called the insertion loss. In the case of active circuits, there is usually gain, or in other words $S_{21}$ is larger than unity. In the case of passive circuits, $S_{12}$ is equal to $S_{21}$ because passive circuits are reciprocal. The only exception is the case of ferrites. In the case of active circuits, $S_{12}$ is different from $S_{21}$ and usually much smaller than unity, since it represents feedback, which is often avoided by design due to the Miller effect. The gain or loss is typically expressed in decibels:

$$\text{gain/insertion loss} = \left|20\log_{10}\left|S_{21}\right|\right| \text{dB} \tag{1.22}$$

## 1.3.2 Noise FOMs

### 1.3.2.1 The noise factor

The previous results actually lead us to a very important and key point regarding noisy devices, that is, the FOM called the noise factor (NF), which characterizes the degradation of the signal-to-noise ratio (SNR) by the device itself.

The noise factor is defined as follows.

DEFINITION 1.1 *The noise factor* (F) *of a circuit is the ratio of the signal-to-noise ratio at the input of the circuit to the signal-to-noise ratio at the output of the circuit:*

$$F = \frac{S_I/N_I}{S_O/N_O} \tag{1.23}$$

*where*

- $S_I$ *is the power of the signal transmitted from the source to the input of the two-port network*
- $S_O$ *is the power of the signal transmitted from the output of the two-port network to the load*
- $N_I$ *is the power of the noise transmitted from the source impedance* $Z_S$ *at temperature* $T_0 = 290\ K$ *to the input of the two-port network*
- $N_O$ *is the power of the noise transmitted from the output of the two-port network to the load*

The noise factor can be expressed as

$$F = \frac{N_{ad} + G_A N_{aI}}{G_A N_{aI}} \tag{1.24}$$

where $G_A$ is the available power gain of the two-port network (for its definition, see Section 1.8), $N_{ad}$ is the additional available noise power generated by the two-port network, and $N_{aI}$ is the available noise power generated by the source impedance:

$$N_{aI} = 4k_B T_0\, \Delta f \tag{1.25}$$

As can be seen from Eq. (1.24), $F$ is always greater than unity, and it does not depend upon the load $Z_L$. It depends exclusively upon the source impedance $Z_S$.

Using reference [3], the noise factor can also be related to the $S$-parameters by:

$$F = F_{min} + 4\frac{R_N}{Z_0}\frac{|\Gamma_{OPT} - \Gamma_s|^2}{(1 - |\Gamma_s|^2)|1 + \Gamma_{OPT}|^2} \tag{1.26}$$

where $F_{min}$ is the minimum noise factor, $R_N$ is called the noise resistance, $\Gamma_{OPT}$ is the optimum source reflection coefficient for which the noise factor is minimum.

This formulation can also be made in terms of $Y$-parameters, and can be expressed as a function of the source admittance $Y_S$:

$$F = F_{\min} + \frac{R_N}{\mathrm{Re}(Y_S)} |Y_S - Y_{OPT}|^2 \quad (1.27)$$

where $Y_{OPT}$ is the optimum source admittance for which the noise factor is minimum.

The terms $F_{MIN}$, $R_N$, and $\Gamma_{OPT}$ (or $Y_{OPT}$) constitute the four noise parameters of the two-port network. They can be related to the correlation matrices very easily [3].

The noise figure (NF) is simply the logarithmic version of the noise factor, $F$.

#### 1.3.2.2 Cascade of noisy two-port components

If we cascade two noisy devices with noise factors $F_1$ and $F_2$, and with available power gains $G_{A1}$ and $G_{A2}$, with a source impedance at temperature $T_0 = 290$ K, the additional available noise powers are

$$\begin{aligned} N_{ad1} &= (F_1 - 1) G_{A1} k_B T_0 \, \Delta f \\ N_{ad2} &= (F_2 - 1) G_{A2} k_B T_0 \, \Delta f \end{aligned} \quad (1.28)$$

The available noise power at the output of the second two-port network is

$$N_{aO2} = k_B T_0 \, \Delta f \; G_{A1} G_{A2} + N_{ad1} G_{A2} + N_{ad2} \quad (1.29)$$

The total noise factor is thus

$$F = \frac{N_{aO2}}{k_B T_0 \, \Delta f \; G_{A1} G_{A2}} \quad (1.30)$$

This finally leads to the well-known noise Friis formula,

$$F = F_1 + \frac{F_2 - 1}{G_{A1}} \quad (1.31)$$

In this expression the gain is actually the available power gain of the first two-port network, which depends on the output impedance of the first network. $F_1$ depends on the source impedance, and $F_2$ depends on the output impedance of the first two-port network.

The general Friis formula is

$$F = F_1 + \frac{F_2 - 1}{G_{A1}} + \frac{F_3 - 1}{G_{A1} G_{A2}} + \cdots + \frac{F_N - 1}{G_{A1} G_{A2} \ldots G_{AN-1}}$$

## 1.4 Nonlinear two-port networks

In order to better understand nonlinear distortion effects, let us start by explaining the fundamental properties of nonlinear systems. Since a nonlinear system is defined as a system that is not linear, we will start by explaining the fundamentals of linear systems.

Linear systems are systems that obey superposition. This means that they are systems whose output to a signal composed by the sum of elementary signals can be given as the sum of the outputs to these elementary signals when taken individually.

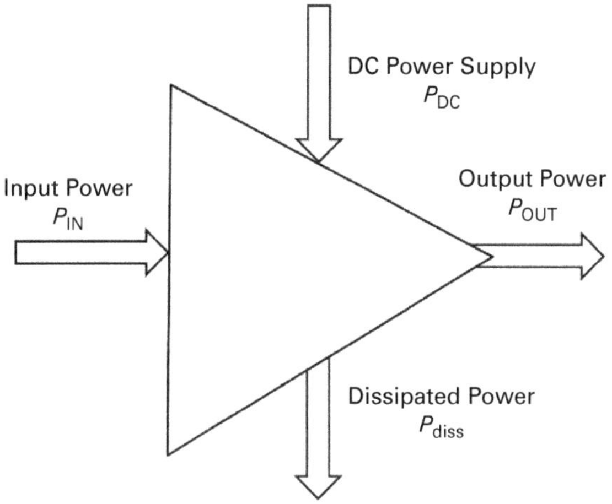


**Figure 1.8** The power balance of a nonlinear system.

This can be stated as

$$y(t) = S_L[x(t)] = k_1 y_1(t) + k_2 y_2(t) \tag{1.32}$$

where $x(t) = k_1 x_1(t) + k_2 x_2(t)$, $y_1(t) = S_L[x_1(t)]$, and $y_2(t) = S_L[x_2(t)]$. Any system that does not obey Eq. (1.32) is said to be a nonlinear system. Actually, this violation of the superposition theorem is the typical rule rather than being the exception. For the remainder of this section, we assume the two-port network to be an amplifier.

To better understand this mechanism, consider the general active system of Fig. 1.8, where $P_{IN}$ and $P_{OUT}$ are the input power entering the amplifier and the output power going to the load, respectively; $P_{DC}$ is the DC power delivered to the amplifier by the power supply; and $P_{diss}$ is the total amount of power lost, by being dissipated in the form of heat or in any other form [4].

Using the definition of operating power gain, $G = P_{OUT}/P_{IN}$ (see also Section 1.8.1), and considering that the fundamental energy-conservation principle requires that $P_L + P_{diss} = P_{IN} + P_{DC}$, we can write the operating power gain as

$$G = 1 + \frac{P_{DC} - P_{diss}}{P_{IN}} \tag{1.33}$$

From this equation we can see that, since $P_{diss}$ has a theoretical minimum of zero and $P_{DC}$ is limited by the finite available power from the supply, the amplifier cannot maintain a constant power gain for increasing input power.

This will lead the amplifier to start to deviate from linearity at a certain input power, and thus start to become nonlinear. Figure 1.9 presents this result by sketching the operating power gain of an amplifier versus the input power rise.

### 1.4.1 Nonlinear generation

In order to evaluate how the inherent nonlinear phenomena can affect amplifiers, let us consider a simple analysis, where we will compare the responses of simple linear and nonlinear systems to typical inputs encountered in wireless technology.

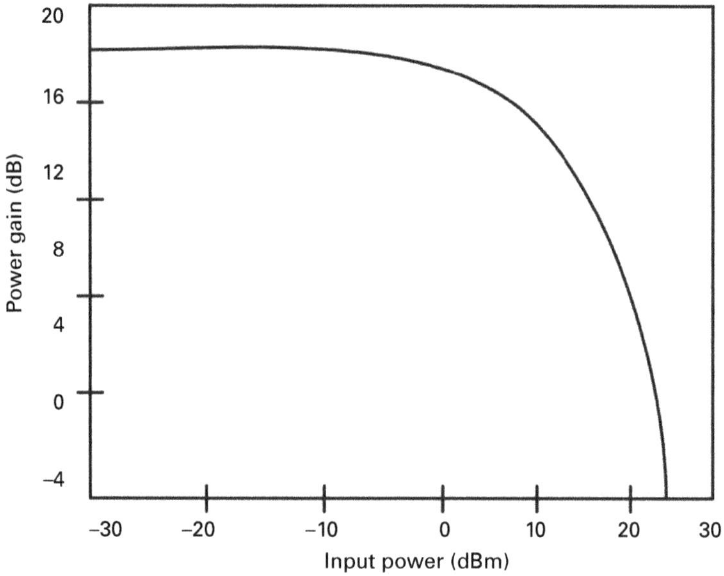


**Figure 1.9** The nonlinear behavior of the variation of power gain versus input power.

In wireless systems, which are mainly based on radio-frequency communications, the stimulus inputs are usually sinusoids, with these being amplitude- and phase-modulated by some baseband information signal. Therefore the input signal of these systems can be written as

$$x(t) = A(t)\cos[\omega_c t + \theta(t)] \tag{1.34}$$

where $A(t)$ is the time-dependent amplitude, $\omega_c$ is the carrier pulsation, and $\phi(t)$ is the modulated phase.

The simplest form of nonlinear behavior that allows us to mathematically describe the response is that in which the nonlinearity is represented by a polynomial [4]:

$$y_{NL}(t) = a_1 x(t-\tau_1) + a_2 x(t-\tau_2)^2 + a_3 x(t-\tau_3)^3 + \cdots \tag{1.35}$$

In this case the polynomial was truncated to the third order to simplify the calculations, but a higher-degree polynomial can obviously be used. Actually, this type of approximation is the first solution to the more complex nonlinear behavior of an amplifier, and, if the $a_s$ coefficients and $\tau_s$ are changed, then many systems can be modeled in this way.

The system actually behaves as a linear one, if $x(t) \gg x(t)^2, x(t)^3$, and $y_{NL}(t) \approx y_L(t) = S_L[x(t)] = a_1 x(t-\tau_1)$.

In this case, the linear response will be

$$y_L(t) = a_1 A(t-\tau_1)\cos[\omega_c t + \theta(t-\tau_1) - \phi_1] \tag{1.36}$$

and the overall nonlinear response can be written as

$$\begin{aligned} y_{\mathrm{NL}}(t) &= a_1 A(t-\tau_1)\cos[\omega_c t + \theta(t-\tau_1) - \phi_1] \\ &+ a_2 A(t-\tau_2)^2 \cos[\omega_c t + \theta(t-\tau_2) - \phi_2]^2 \\ &+ a_3 A(t-\tau_3)^3 \cos[\omega_c t + \theta(t-\tau_3) - \phi_3]^3 \end{aligned} \tag{1.37}$$

Using some trigonometric relations, as presented in [4], the equations can be written as

$$\begin{aligned} y_{\mathrm{NL}}(t) &= a_1 A(t-\tau_1)\cos[\omega_c t + \theta(t-\tau_1) - \phi_1] \\ &+ \frac{1}{2} a_2 A(t-\tau_2)^2 \\ &+ \frac{1}{2} a_2 A(t-\tau_2)^2 \cos[2\omega_c t + 2\theta(t-\tau_2) - 2\phi_2] \\ &+ \frac{3}{4} a_3 A(t-\tau_3)^3 \cos[\omega_c t + \theta(t-\tau_3) - \phi_3] \\ &+ \frac{1}{4} a_3 A(t-\tau_3)^3 \cos[3\omega_c t + 3\theta(t-\tau_3) - 3\phi_3] \end{aligned} \tag{1.38}$$

where $\phi_1 = \omega_c \tau_1$, $\phi_2 = \omega_c \tau_2$, and $\phi_3 = \omega_c \tau_3$.

In typical wireless communication systems, the variation of the modulated signals is usually slow compared with that of the RF carrier, and thus, if the system does not exhibit memory effects, one can write

$$y_{\mathrm{L}}(t) = a_1 A(t)\cos[\omega_c t + \theta(t) - \phi_1] \tag{1.39}$$

and

$$\begin{aligned} y_{\mathrm{NL}}(t) &= a_1 A(t)\cos[\omega_c t + \theta(t) - \phi_1] \\ &+ \frac{1}{2} a_2 A(t)^2 \\ &+ \frac{1}{2} a_2 A(t)^2 \cos[2\omega_c t + 2\theta(t) - 2\phi_2] \\ &+ \frac{3}{4} a_3 A(t)^3 \cos[\omega_c t + \theta(t) - \phi_3] \\ &+ \frac{1}{4} a_3 A(t)^3 \cos[3\omega_c t + 3\theta(t) - 3\phi_3] \end{aligned} \tag{1.40}$$

From Eqs. (1.39) and (1.40) it is clear that the linear and nonlinear responses are significantly different. For instance the number of terms in the nonlinear formulation is quite high compared with the number in the linear formulation.

Moreover, the output of the linear response is a version of the input signal, with the same spectral contents, but with a variation in amplitude and phase as compared with the input, whereas the nonlinear response consists of a panoply of other spectral components, usually called spectral regrowth. Actually, this is one of the properties of nonlinear systems, namely that, in contrast to a linear system, which can only introduce quantitative changes to the signal spectra, nonlinear systems can qualitatively modify spectra, insofar as they eliminate certain spectral components and generate new ones.

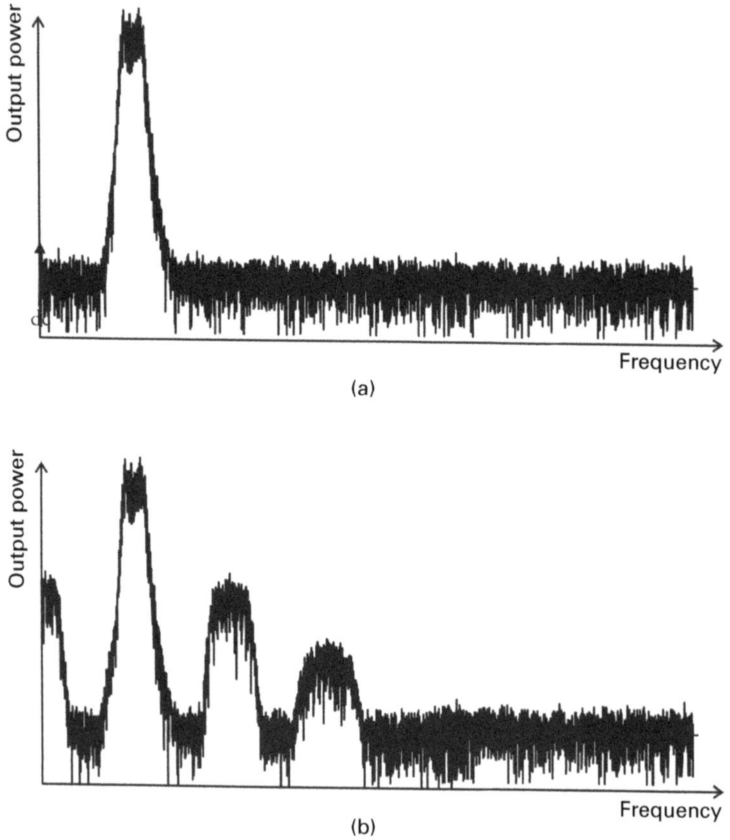


**Figure 1.10** Signal probing throughout the wireless system path: (a) the analog signal at point 4; and (b) the signal at point 5, where the generation of nonlinear distortion is visible.

One example of a typical linear component is a filter, since the output of a filter can in principle change exclusively the amplitude and phase of the input signal, while an example of a nonlinearity is a frequency multiplier, where the output spectrum is completely different from the input spectrum.

In a typical wireless system such as the one presented in Fig. 1.1, the most important source of nonlinear distortion is the power amplifier (PA), but all the components can behave nonlinearly, depending on the input signal excitation.

### 1.4.2 Nonlinear impact in wireless systems

In the previous section the nonlinear generation mechanism was explained using a simple polynomial. In this section we will probe the signal throughout the system presented in Fig. 1.1.

Figure 1.10 presents the spectral content in each of the stages of the wireless system, and allows one to see how the signal changes on traversing the communication path.

As can be seen from the images, the starting signal is nothing more than a bit stream arriving at our digital-to-analog converter, point 1 in Fig. 1.1. Then this signal is further converted to analog and filtered out (point 2), up-converted to an IF channel (point 3), amplified and further up-converted to RF (point 4), amplified again (point 5), and transmitted over the air (point 6).

The signal then traverses the air interface and, at the receiver, it is first filtered out (point 7), then amplified using a low-noise amplifier (point 8), and then down-converted to IF (point 9), and to baseband again (point 10), and reconverted to a digital version (point 11).

Since in this section we are looking mainly at the nonlinear behavior of the RF signal, let us concentrate on points 4 and 5. In this case the input signal has a certain spectral shape, as can be seen in Fig. 1.10(a), and, after the nonlinear behavior of the PA, it appears completely different at point 5, where several clusters of spectra appear, Fig. 1.10(b).

The first cluster is centered at DC and, in practical systems, it consists of two forms of distortion, namely the DC value itself and a cluster of very-low-frequency spectral components centered at DC. The DC value distortion manifests itself as a shift in bias from the quiescent point (defined as the bias point measured without any excitation) to the actual bias point measured when the system is driven at its rated input excitation power.

If we look back at Eq. (1.40), we can understand that the DC component comes from all possible mixing, beat, or nonlinear distortion products of the form $\cos(\omega_i t)\ \cos(\omega_j t)$, whose frequency mixing appears at $\omega_x = \omega_i - \omega_j$, where $\omega_i = \omega_j$.

The low-frequency cluster near DC constitutes a distorted version of the amplitude-modulating information, $A(t)$, as if the input signal had been demodulated. This cluster is, therefore, called the baseband component of the output. In spectral terms their frequency lines are also generated from mixing products at $\omega_x = \omega_i - \omega_j$, but now where $\omega_i \neq \omega_j$.

From Fig. 1.10(b) it is clear that there are some other clusters appearing at $2\omega_c$ and $3\omega_c$. These are the well-known second- and third-harmonic components, usually called the harmonic distortion. Actually, they are high-frequency replicas of the modulated signal.

The cluster appearing at $2\omega_c$ is again generated from all possible mixing products of the form $\cos(\omega_i t)\ \cos(\omega_j t)$, but now the outputs are located at $\omega_x = \omega_i + \omega_j$, where $\omega_i = \omega_j$ $(\omega_x = 2\omega_i = 2\omega_j)$ or $\omega_i \neq \omega_j$.

The third-harmonic cluster appears from all possible mixing products of the form $\cos(\omega_i t)\cos(\omega_j t)\cos(\omega_k t)$, whose outputs are located at $\omega_x = \omega_i + \omega_j + \omega_k$, where $\omega_i = \omega_j = \omega_k$ $(\omega_x = 3\omega_i = 3\omega_j = 3\omega_k)$ or $\omega_i = \omega_j \neq \omega_k$ $(\omega_x = 2\omega_i + \omega_k = 2\omega_j + \omega_k)$ or even $\omega_i \neq \omega_j \neq \omega_k$.

The last-mentioned cluster is the one appearing around $\omega_c$. In this scenario the nonlinear distortion appears near the spectral components of the input signal, but is also exactly coincident with them, and thus is indistinguishable from them.

Unfortunately, and in contrast to the baseband or harmonic distortion, which falls on out-of-band spectral components, and thus could be simply eliminated by bandpass filtering, some of these new in-band distortion components are unaffected by any linear operator that, naturally, must preserve the fundamental components. Thus, they constitute the most important form of distortion in bandpass microwave and wireless subsystems. Since this is actually the most important form of nonlinear distortion in narrowband systems, it is sometimes just called "distortion."

In order to clearly understand and identify the in-band-distortion spectral components, they must be first separated in to the spectral lines that fall exactly over the original ones and the lines that constitute distortion sidebands. In wireless systems, the former are known as *co-channel distortion* and the latter as *adjacent-channel distortion*.

Looking back at our formulation Eq. (1.40), all in-band-distortion products share the form of $\cos(\omega_i t)\cos(\omega_j t)\cos(\omega_k t)$, which is similar to the ones appearing at the third harmonic, but now the spectral outputs are located at $\omega_x = \omega_i + \omega_j - \omega_k$. In this case, and despite the fact that both co-channel and adjacent-channel distortion can be generated by mixing products obeying $\omega_i = \omega_j \neq \omega_k$ $(\omega_x = 2\omega_i - \omega_k = 2\omega_j - \omega_k)$ or $\omega_i \neq \omega_j \neq \omega_k$, only the mixing terms obeying $\omega_i = \omega_j = \omega_k$ $(\omega_x = \omega_i)$ or $\omega_i \neq \omega_j = \omega_k$ $(\omega_x = \omega_i)$ fall on top of the co-channel distortion.

## 1.5 Nonlinear FOMs

Let us now try to identify how we can account for nonlinearity in wireless two-port networks. To this end we will use different signal excitations, since those will reveal different aspects of the nonlinear behavior. We will start first with a single-tone excitation, and then proceed to the best-known signal excitation for nonlinear distortion, namely two-tone excitation, and then the multi-sine excitation figures of merit will also be addressed. Finally, a real modulated wireless signal will be used to define the most important FOMs in wireless systems.

### 1.5.1 Nonlinear single-tone FOMs

We start by considering that $x(t)$ in Eq. (1.35) is a single sinusoid, $x(t) = A\cos(\omega_c t)$. The output signal is described by Eq. (1.40), and, if we consider that the input signal does not have a phase delay, it can be further simplified to

$$\begin{aligned} y_{NL}(t) = {} & a_1 A\cos[\omega_c t - \phi_1] \\ & + \frac{1}{2}a_2A^2 \\ & + \frac{1}{2}a_2A^2\cos[2\omega_c t - 2\phi_2] \\ & + \frac{3}{4}a_3A^3\cos[\omega_c t - \phi_3] \\ & + \frac{1}{4}a_3A^3\cos[3\omega_c t - 3\phi_3] \end{aligned} \tag{1.41}$$

In this case the output consists of single-tone spectral components appearing at DC, in the same frequency component as the input signal, and at the second and third harmonics.

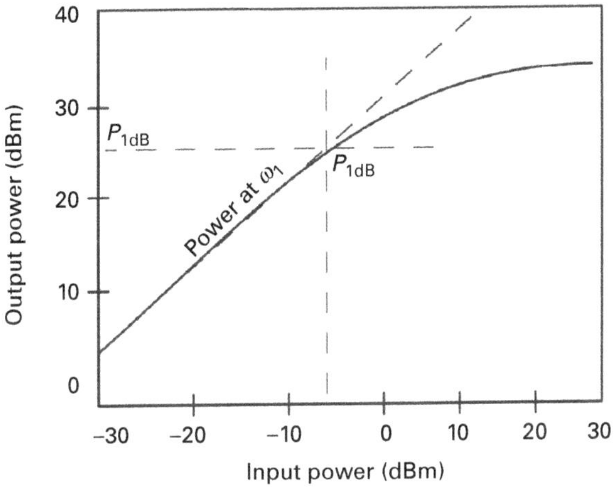


**Figure 1.11** AM–AM curves, where the output power is plotted versus the input power increase. The 1-dB-compression point is also visible in the image.

Actually, the output amplitude and phase variation versus input drive manifest themselves as if the nonlinear device could convert input amplitude variations into output amplitude and phase changes or, in other words, as if it could transform possible amplitude modulation (AM) associated with its input into output amplitude modulation (AM–AM conversion) or phase modulation (AM–PM conversion).

AM–AM conversion is particularly important in systems that are based on amplitude modulation, while AM–PM has its major impact in modern wireless telecommunication systems that rely on phase-modulation formats.

If a careful analysis is done at the harmonics, we can also calculate the ratio of the integrated power of all the harmonics to the measured power at the fundamental, a figure of merit named total harmonic distortion (THD).

#### 1.5.1.1 AM–AM

The AM–AM figure of merit describes the relationship between the output amplitude and the input amplitude at the fundamental frequency [4].

Figure 1.11 presents the AM–AM characteristic, where the power of each of the fundamental spectral components is plotted versus its input counterpart. As can be seen, it characterizes the gain compression or expansion of a nonlinear device versus the input drive level.

One of the most important FOMs that can be extracted from this type of characterization is called the 1-dB-compression point, $P_{1dB}$.

#### 1.5.1.2 The 1-dB-compression point ($P_{1dB}$)

DEFINITION 1.2 *The 1-dB-compression point ($P_{1dB}$) is defined as the output power level at which the signal output is compressed by 1 dB, compared with the output power level that would be obtained by simply extrapolating the linear system's small-signal characteristic.*

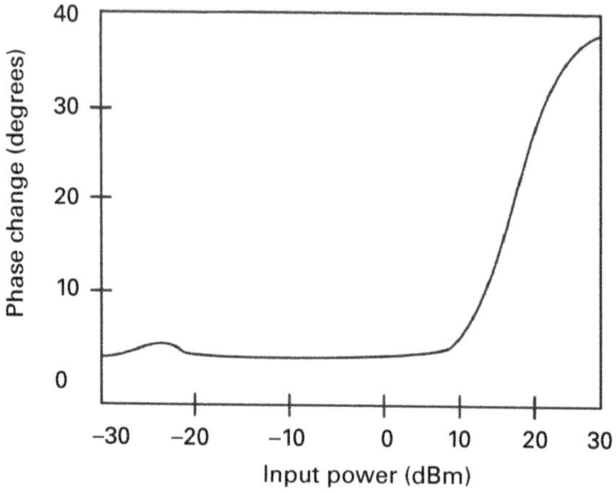


**Figure 1.12** AM–PM curves, where the phase delay of the output signal is visible when the input power is varied.

Thus, the $P_{1dB}$ FOM also corresponds to a 1-dB gain deviation from its small-signal value, as depicted in Fig. 1.9 and Fig. 1.11.

### 1.5.1.3 AM–PM

Since the co-channel nonlinear distortion actually falls on top of the input signal spectra, Eq. (1.41), the resulting output component at that frequency will be the addition of two vectors: the linear output signal, plus a version of the nonlinear distortion. So vector addition can also determine a phase variation of the resultant output, when the input level varies, as shown in Fig. 1.12.

The change of the output signal phase, $\phi(\omega, A_i)$, with increasing input power is the AM–PM characteristic and may be expressed as a certain phase deviation, in degrees/dB, at a predetermined input power.

### 1.5.1.4 Total harmonic distortion

The final FOM in connection with single-tone excitation is one that accounts for the higher-order harmonics, and it is called total harmonic distortion (THD).

DEFINITION 1.3 *The total harmonic distortion (THD) is defined as the ratio between the square roots of the total harmonic output power and the output power at the fundamental frequency.*

Therefore the THD can be expressed as

$$\mathrm{THD} = \frac{\sqrt{1/T \int_0^T \left[\sum_{r=2}^{\infty} A_{0_r}(\omega, A_i)\cos[r\omega t + \theta_{0_r}(\omega, A_i)]\right]^2 dt}}{\sqrt{1/T \int_0^T \left[A_{0_1}(\omega, A_i)\cos[\omega t + \theta_{0_1}(\omega, A_i)]\right]^2 dt}} \tag{1.42}$$

and, for the polynomial case, we will have

$$\mathrm{THD} = \frac{\sqrt{\frac{1}{8}a_2^2 A_i^4 + (1/32)a_3^2 A_i^6 + \cdots}}{\sqrt{a_1^2 A_i^2/2}} \tag{1.43}$$

## 1.5.2 Nonlinear two-tone FOMs

A single-tone signal unfortunately is just a first approach to the characterization of a nonlinear two-port network. Actually, as was seen previously, the single-tone signal can be used to evaluate the gain compression and expansion, and harmonic generation, but no information is given about the bandwidth of the signal, or about the distortion appearing in-band.

In order to get a better insight into these in-band-distortion products, RF engineers started to use so-called two-tone excitation signals.

A two-tone signal is composed of a summation of two sinusoidal signals,

$$x(t) = A_1 \cos(\omega_1 t) + A_2 \cos(\omega_2 t) \tag{1.44}$$

Since the input is now composed of two different carriers, many more mixing products will be generated when it traverses the polynomial presented in Eq. (1.35). Therefore, it is convenient to count all of them in a systematic manner. Hence the sine representation will be substituted by its Euler expansion representation:

$$\begin{aligned} x(t) &= A_1 \cos(\omega_1 t) + A_2 \cos(\omega_2 t) \\ &= A_1 \frac{e^{j[\omega_1 t]} + e^{-j[\omega_1 t]}}{2} + A_2 \frac{e^{j[\omega_2 t]} + e^{-j[\omega_2 t]}}{2} \end{aligned} \tag{1.45}$$

This type of formulation actually allows us to calculate all the mixing products arising from the polynomial calculations, since the input can now be viewed as the sum of four terms, each one involving a different frequency. That is, we are assuming that each sinusoidal function involves a positive- and a negative frequency component (i.e., the corresponding positive and negative sides of the Fourier spectrum), so that any combination of tones can be represented as

$$\begin{aligned} x(t) &= \sum_{r=1}^{R} A_r \cos(\omega_r t) \\ &= \frac{1}{2} \sum_{r=-R;r\neq 0}^{R} A_r e^{j\omega_r t} \end{aligned} \tag{1.46}$$

where $r \neq 0$, and $A_r = A_{-r}^*$ for real signals.

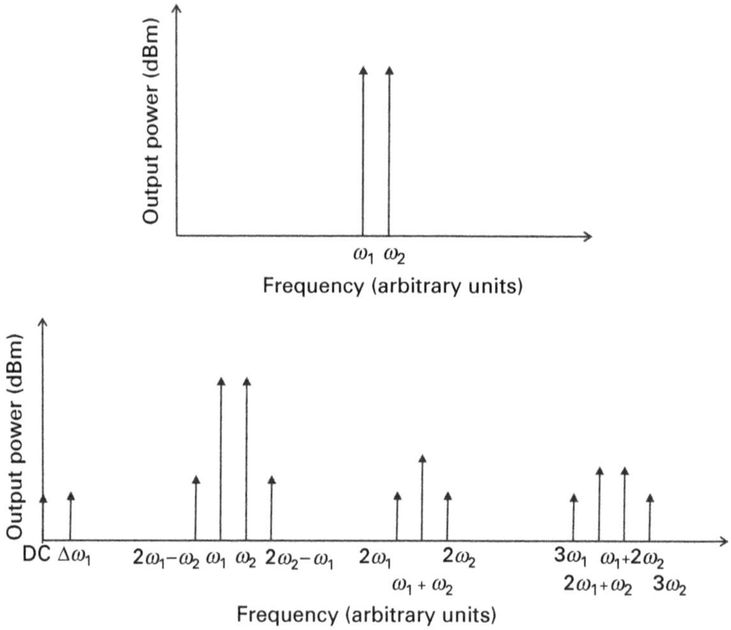


**Figure 1.13** The spectrum arrangement of a two-tone signal traversing a nonlinearity, where different spectrum clusters can be seen.

The output of the polynomial model for this type of formulation is now much simpler to develop, and for each mixing value we will have

$$
\begin{aligned}
y_{\mathrm{NL}n}(t) &= \frac{1}{2^n} a_n \left[ \sum_{r=-R}^{R} A_r e^{j\omega_r t} \right]^n \\
&= \frac{1}{2^n} a_n \sum_{r_1=-R}^{R} \cdots \sum_{r_n=-R}^{R} A_{r_1} \cdots A_{r_n} e^{j(\omega_{r_1}+\cdots+\omega_{r_n})t}
\end{aligned}
\tag{1.47}
$$

The frequency components arising from this type of mixing are all possible combinations of the input $\omega_r$:

$$
\begin{aligned}
\omega_n &= \omega_{r_1} + \cdots + \omega_{r_n} \\
&= m_{-R}\omega_{-R} + \cdots + m_{-1}\omega_{-1} + m_1\omega_1 + \cdots + m_R\omega_R
\end{aligned}
\tag{1.48}
$$

where the vector $[m_{-R} \ldots m_{-1} m_1 \ldots m_R]$ is the $n$th order mixing vector, which must satisfy

$$
\begin{aligned}
\sum_{r=-R}^{R} m_r &= m_{-R} + \cdots + m_{-1} + m_1 + \cdots + m_R \\
&= n
\end{aligned}
\tag{1.49}
$$

In the case of a two-tone signal (Fig. 1.13), the first-order mixing, arising from the linear response (coefficient $a_1$ in the polynomial), will be

$$-\omega_2, -\omega_1, \omega_1, \omega_2 \tag{1.50}$$

Adding then the second order (coefficient $a_2$), we will have

$$-2\omega_2, -\omega_2 - \omega_1, -2\omega_1, \omega_1 - \omega_2, \mathrm{DC}, \omega_2 - \omega_1, 2\omega_1, \omega_1 + \omega_2, 2\omega_2 \tag{1.51}$$

and, for the third-order coefficient $a_3$,

$$\begin{aligned}
&-3\omega_2, -2\omega_2 - \omega_1, -\omega_2 - 2\omega_1, -3\omega_1, \\
&-2\omega_2 + \omega_1, -\omega_2, -\omega_1, -2\omega_1 + \omega_2, \\
&\ 2\omega_1 - \omega_2, \omega_1, \omega_2, 2\omega_2 - \omega_1, \\
&\ 3\omega_1, 2\omega_1 + \omega_2, \omega_1 + 2\omega_2, 3\omega_2
\end{aligned} \tag{1.52}$$

Obviously there are several ways in which these mixing products can be gathered, and the reader can find the calculations in [4, 5]. If the Euler coefficients corresponding to each mixing product are now added, we obtain the following expression for the output of our nonlinearity when excited by a two-tone signal:

$$\begin{aligned}
y_{\mathrm{NL}}(t) = {} & a_1 A_1 \cos(\omega_1 t - \phi_{1_{10}}) + a_1 A_2 \cos(\omega_2 t - \phi_{1_{10}}) \\
& + \frac{1}{2} a_2 (A_1^2 + A_2^2) \\
& + a_2 A_1 A_2 \cos[(\omega_2 - \omega_1)t - \phi_{2_{-11}}] \\
& + a_2 A_1 A_2 \cos[(\omega_1 + \omega_2)t - \phi_{2_{11}}] \\
& + \frac{1}{2} a_2 A_1^2 \cos(2\omega_1 t - \phi_{2_{20}}) + \frac{1}{2} a_2 A_2^2 \cos(2\omega_2 t - \phi_{2_{02}}) \\
& + \frac{3}{4} a_3 A_1^2 A_2 \cos[(2\omega_1 - \omega_2)t - \phi_{3_{2-1}}] \\
& + \left(\frac{3}{4} a_3 A_1^3 + \frac{6}{4} a_3 A_1 A_2^2\right) \cos(\omega_1 t - \phi_{3_{10}}) \\
& + \left(\frac{3}{4} a_3 A_2^3 + \frac{6}{4} a_3 A_2 A_1^2\right) \cos(\omega_2 t - \phi_{3_{01}}) \\
& + \frac{3}{4} a_3 A_1 A_2^2 \cos[(2\omega_2 - \omega_1)t - \phi_{3_{-12}}] \\
& + \frac{1}{4} a_3 A_1^3 \cos(3\omega_1 t - \phi_{3_{30}}) \\
& + \frac{3}{4} a_3 A_1^2 A_2 \cos[(2\omega_1 + \omega_2)t - \phi_{3_{21}}] \\
& + \frac{3}{4} a_3 A_1 A_2^2 \cos[(\omega_1 + 2\omega_2)t - \phi_{3_{12}}] \\
& + \frac{1}{4} a_3 A_2^3 \cos(3\omega_2 t - \phi_{3_{03}})
\end{aligned} \tag{1.53}$$

where $\phi_{1_{10}} = \omega_1\tau_1$, $\phi_{1_{01}} = \omega_2\tau_1$, $\phi_{2_{-11}} = \omega_2\tau_2 - \omega_1\tau_2$, $\phi_{2_{20}} = 2\omega_1\tau_2$, $\phi_{2_{11}} = \omega_1\tau_2 + \omega_2\tau_2$, $\phi_{2_{02}} = 2\omega_2\tau_2$, $\phi_{3_{2-1}} = 2\omega_1\tau_3 + \omega_2\tau_3$, $\phi_{3_{10}} = \omega_1\tau_3$, $\phi_{3_{01}} = \omega_2\tau_3$, $\phi_{3_{-12}} = 2\omega_2\tau_3 - \omega_1\tau_3$, $\phi_{3_{30}} = 3\omega_1\tau_3$, $\phi_{3_{21}} = 2\omega_1\tau_3 + \omega_2\tau_3$, $\phi_{3_{12}} = \omega_1\tau_3 + 2\omega_2\tau_3$, and $\phi_{3_{03}} = 3\omega_2\tau_3$.

If we look exclusively at the in-band distortion, the output components will be

$$\begin{aligned} y_{NL_{\text{in-band}}}(t) &= a_1 A_1 \cos\left(\omega_1 t - \phi_{1_{10}}\right) + a_1 A_2 \cos\left(\omega_2 t - \phi_{1_{10}}\right) \\ &+ \frac{3}{4} a_3 A_1^2 A_2 \cos\left[(2\omega_1 - \omega_2)t - \phi_{3_{2-1}}\right] \\ &+ \left(\frac{3}{4} a_3 A_1^3 + \frac{6}{4} a_3 A_1 A_2^2\right) \cos\left(\omega_1 t - \phi_{3_{10}}\right) \\ &+ \left(\frac{3}{4} a_3 A_2^3 + \frac{6}{4} a_3 A_2 A_1^2\right) \cos\left(\omega_2 t - \phi_{3_{01}}\right) \\ &+ \frac{3}{4} a_3 A_1 A_2^2 \cos\left[(2\omega_2 - \omega_1)t - \phi_{3_{-12}}\right] \end{aligned} \tag{1.54}$$

From this equation it is clear that the in-band distortion in this case is much richer than that in the single-sinusoid case. With the two-tone excitation one can identify the linear components arising from the $a_1$ terms and the nonlinear components arising from the $a_3$ terms.

For the nonlinear components, two further distinctions can be made, since two terms will fall in frequency sidebands, namely the cases of $2\omega_1 - \omega_2$ and $2\omega_2 - \omega_1$, and two other terms will fall right on top of the input signal at $\omega_1$ and $\omega_2$.

The terms falling in the sidebands are normally called intermodulation distortion (IMD). Actually, every nonlinear mixing product can be denominated as an intermodulation component since it results from intermodulating two or more different tones. But, although it cannot be said to be universal practice, the term IMD is usually reserved for those particular sideband components.

This form of distortion actually constitutes a form of adjacent-channel distortion.

The terms that actually fall on top of $\omega_1$ and $\omega_2$ are known as the co-channel distortion, and in fact, if we look carefully, we see that they can actually be divided into two separate forms. For instance, for $\omega_1$,

$$\begin{aligned} y_{NL_{\text{co-channel}}}(t) &= \left(\frac{3}{4} a_3 A_1^3 + \frac{6}{4} a_3 A_1 A_2^2\right) \cos\left(\omega_1 t - \phi_{3_{10}}\right) \\ &= \frac{3}{4} a_3 A_1^3 \cos\left(\omega_1 t - \phi_{3_{10}}\right) + \frac{6}{4} a_3 A_1 A_2^2 \cos\left(\omega_1 t - \phi_{3_{10}}\right) \end{aligned} \tag{1.55}$$

which corresponds to a term that depends only on $A_1^3$, which is perfectly correlated with the input signal at $A_1$, and another term that falls on top of $\omega_1$ but depends also on the $A_2$ term, meaning that it can be uncorrelated with the input signal.

Actually, the correlated version of the output signal $\frac{3}{4} a_3 A_1^3 \cos(\omega_1 t - \phi_{3_{10}})$ is the term that is responsible for the compression or expansion of the device gain. This is similar to what was previously said regarding single-tone excitations, which we called AM–AM and AM–PM responses.

The other term, $\frac{6}{4} a_3 A_1 A_2^2 \cos(\omega_1 t - \phi_{3_{10}})$, which also includes a contribution from $A_2$ and can be uncorrelated with the input signal, is actually the worst problem in terms of communication signals. It is sometimes referred to as distortion noise. In wireless communications it is this type of nonlinear distortion that can degrade, for instance, the

**Table 1.1** Two-tone nonlinear distortion mixing products up to third order

| Mixing product frequency | Output amplitude | Result |
|---|---|---|
| $\omega_1$ | $(1/2)a_1 A_1$ | Linear response |
| $\omega_2$ | $(1/2)a_1 A_2$ | Linear response |
| $\omega_1 - \omega_1$ | $(1/2)a_2 A_1^2$ | Change in DC bias point |
| $\omega_2 - \omega_2$ | $(1/2)a_2 A_2^2$ | Change in DC bias point |
| $\omega_2 - \omega_1$ | $(1/2)a_2 A_1 A_2$ | Second-order mixing response |
| $2\omega_1$ | $(1/4)a_2 A_1^2$ | Second-order harmonic response |
| $\omega_1 + \omega_2$ | $(1/2)a_2 A_1 A_2$ | Second-order mixing response |
| $2\omega_2$ | $(1/4)a_2 A_2^2$ | Second-order harmonic response |
| $2\omega_1 - \omega_2$ | $(3/8)a_3 A_1^2 A_2$ | Third-order intermodulation distortion |
| $\omega_1 + \omega_2 - \omega_2$ | $(3/4)a_3 A_1 A_2^2$ | Cross-modulation response |
| $\omega_1 + \omega_1 - \omega_1$ | $(3/8)a_3 A_1^3$ | AM–AM and AM–PM response |
| $\omega_2 + \omega_2 - \omega_2$ | $(3/8)a_3 A_2^3$ | AM–AM and AM–PM response |
| $\omega_2 + \omega_1 - \omega_1$ | $(3/4)a_3 A_1^2 A_1^2$ | cross-modulation response |
| $2\omega_2 - \omega_1$ | $(3/8)a_3 A_1 A_2^2$ | Third-order intermodulation distortion |
| $3\omega_1$ | $(1/8)a_3 A_1^3$ | Third-order harmonic response |
| $2\omega_1 + \omega_2$ | $(3/8)a_3 A_1^2 A_2$ | Third-order mixing response |
| $2\omega_2 + \omega_1$ | $(3/8)a_3 A_1 A_2^2$ | Third-order mixing response |
| $3\omega_2$ | $(1/8)a_3 A_2^3$ | Third-order harmonic response |

error-vector magnitude (the definition of which will be given later) in digital communication standards.

Table 1.1 summarizes the above definitions by identifying all of the distortion components falling on the positive side of the spectrum which are present in the output of our third-degree polynomial subjected to a two-tone excitation signal.

Following this nonlinear study for a two-tone signal excitation, some figures of merit can be defined.

#### 1.5.2.1 The intermodulation ratio

The intermodulation ratio (IMR) is, as the name states, the ratio between the power corresponding to the output that appears exactly at the same positions as the input spectral components (these components will be called from now on the power at the fundamental frequency) and the power corresponding to the intermodulation power.

DEFINITION 1.4 *The intermodulation ratio (IMR) is defined as the ratio between the fundamental and intermodulation (IMD) output powers:*

$$\mathrm{IMR} = \frac{P_{\mathrm{out_{fund}}}}{P_{\mathrm{IMD}}} \tag{1.56}$$

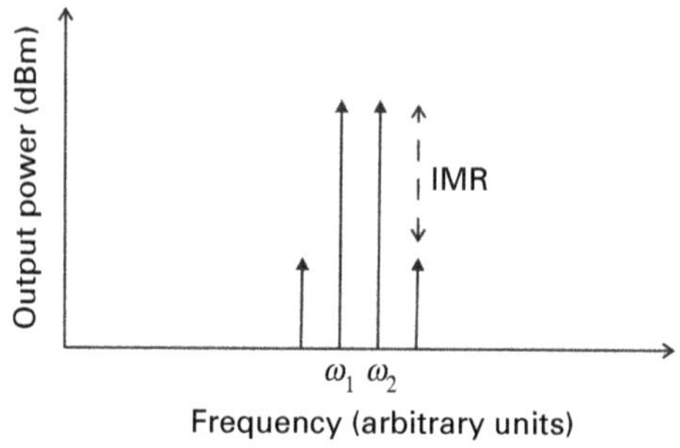


**Figure 1.14** IMR definition in a two-tone excitation.

It should be noticed here that the output power at the fundamental frequency already includes some nonlinear distortion that appears at the same frequency as the input. This was seen in Section 1.5.2 as the co-channel distortion responsible for the AM–AM curves for the case of single-tone excitation, and consequently for the gain compression and expansion. On considering Fig. 1.14 and Eq. (1.56), it is clear that the intermodulation ratio refers only to the in-band nonlinear distortion, not to the harmonic content. This measure is usually described in dBc, meaning decibels below carrier. It should also be pointed out here that the upper IMD and the lower IMD may be different, which is called IMD asymmetry [6]. The IMR must then be defined as upper or lower.

In order to observe the results for the two-tone case, as was stated above, the in-band distortion is

$$
\begin{aligned}
y_{\mathrm{NL}_{\text{in-band}}}(t) = {} & a_1 A_1 \cos\left(\omega_1 t - \phi_{1_{10}}\right) + a_1 A_2 \cos\left(\omega_2 t - \phi_{1_{10}}\right) \\
& + \frac{3}{4} a_3 A_1^2 A_2 \cos\left[(2\omega_1 - \omega_2)t - \phi_{3_{2-1}}\right] \\
& + \left(\frac{3}{4} a_3 A_1^3 + \frac{6}{4} a_3 A_1 A_2^2\right) \cos\left(\omega_1 t - \phi_{3_{10}}\right) \\
& + \left(\frac{3}{4} a_3 A_2^3 + \frac{6}{4} a_3 A_2 A_1^2\right) \cos\left(\omega_2 t - \phi_{3_{01}}\right) \\
& + \frac{3}{4} a_3 A_1 A_2^2 \cos\left[(2\omega_2 - \omega_1)t - \phi_{3_{-12}}\right]
\end{aligned}
\tag{1.57}
$$

which has terms that are clearly co-channel distortion, and thus will add to the output linear response, namely those appearing at

$$
\begin{aligned}
y_{\mathrm{NL}_{\text{in-band-co-channel}}}(t) = {} & a_1 A_{i1} \cos\left(\omega_1 t - \phi_{1_{10}}\right) + a_1 A_{i2} \cos\left(\omega_2 t - \phi_{1_{10}}\right) \\
& + \left(\frac{3}{4} a_3 A_{i1}^3 + \frac{6}{4} a_3 A_{i1} A_{i2}^2\right) \cos\left(\omega_1 t - \phi_{3_{10}}\right) \\
& + \left(\frac{3}{4} a_3 A_{i2}^3 + \frac{6}{4} a_3 A_{i2} A_{i1}^2\right) \cos\left(\omega_2 t - \phi_{3_{01}}\right)
\end{aligned}
\tag{1.58}
$$

In this case the IMD power and the fundamental linear output power will be

$$P_{\mathrm{IMD}}(2\omega_1-\omega_2)=\frac{1}{T_{2\omega_1-\omega_2}}\int_0^{2\omega_1-\omega_2}\left[\frac{3}{4}a_3A_{i1}^2A_{i2}\cos\left[(2\omega_1-\omega_2)t-\phi_{3_{2-1}}\right]\right]^2dt$$
$$=\frac{9}{32}a_3^2A_{i1}^4A_{i2}^2$$
$$P_{\mathrm{fund}}(\omega_1)=\frac{1}{T_{\omega_1}}\int_0^{\omega_1}\left[a_1A_{i1}\cos\left(\omega_1t-\phi_{1_{10}}\right)\right.$$
$$\left.+\left(\frac{3}{4}a_3A_{i1}^3+\frac{6}{4}a_3A_{i1}A_{i2}^2\right)\cos\left(\omega_1t-\phi_{3_{10}}\right)\right]^2dt$$
$$=\frac{1}{2}a_1^2A_{i1}^2+\frac{1}{2}\left(\frac{3}{4}a_3A_{i1}^3+\frac{6}{4}a_3A_{i1}A_{i2}^2\right)^2$$
$$+a_1A_{i1}\left(\frac{3}{4}a_3A_{i1}^3+\frac{6}{4}a_3A_{i1}A_{i2}^2\right)\cos\left(\phi_{1_{10}}-\phi_{3_{10}}\right) \tag{1.59}$$

The same calculations should be done for the IMD at $2\omega_2-\omega_1$, resulting in an IMR for two tones as

$$\mathrm{IMR}_{2\mathrm{t_{low}}}=\frac{32}{9a_3^2A_{i1}^4A_{i2}^2}$$
$$\times\left[\frac{1}{2}a_1^2A_{i1}^2+\frac{1}{2}\left(\frac{3}{4}a_3A_{i1}^3+\frac{6}{4}a_3A_{i1}A_{i2}^2\right)^2\right.$$
$$\left.+a_1A_{i1}\left(\frac{3}{4}a_3A_{i1}^3+\frac{6}{4}a_3A_{i1}A_{i2}^2\right)\cos(\phi_{1_{10}}-\phi_{3_{10}})\right] \tag{1.60}$$

Nevertheless, for low-power signals the nonlinear contribution to the co-channel distortion is insignificant, and thus sometimes only the linear output power is considered when calculating the overall IMR, which will be

$$\mathrm{IMR}_{2\mathrm{t_{low}}}=\frac{\frac{1}{2}a_1^2A_{i1}^2}{(9/32)a_3^2A_{i1}^4A_{i2}^2} \tag{1.61}$$

For equal input signal amplitude in both tones, $A_i=A_{i1}=A_{i2}$, the IMR value will finally be

$$\mathrm{IMR}_{2\mathrm{t}}=\frac{\frac{1}{2}a_1^2A_i^2}{(9/32)a_3^2A_i^6}$$
$$=\frac{16a_1^2}{9a_3^2A_i^4} \tag{1.62}$$

### 1.5.2.2 Underlying linear gain

Actually it should also be mentioned here that sometimes a figure of merit called underlying linear gain (ULG) is defined. This gain accounts for the overall output signal that is correlated with the input signal, thus the ULG is given by

$$\mathrm{ULG}=\frac{P_{\mathrm{out_{fund}}}}{P_{\mathrm{in_{fund}}}} \tag{1.63}$$

For a two-tone excitation it will be

$$\mathrm{ULG}_{2\mathrm{t}_{\mathrm{low}}} = \frac{\frac{1}{2}a_1^2 A_{i1}^2 + \frac{1}{2}\left(\frac{3}{4}a_3 A_{i1}^3\right)^2 + a_1 A_{i1}\left(\frac{3}{4}a_3 A_{i1}^3\right)\cos\left(\phi_{1_{10}} - \phi_{3_{10}}\right)}{\frac{1}{2}A_{i1}^2} \tag{1.64}$$

In the case of linear systems, this gain reduces to the linear gain, that is LG $= \frac{1}{2}a_1^2 A_{i1}^2 / \left(\frac{1}{2}A_{i1}^2\right) = a_1^2$. Actually, with the ULG we are accounting for the AM–AM impact on the overall gain.

#### 1.5.2.3 Intercept points

If the output power at the fundamental and that at the IMD spectral components are plotted versus input power for traditional nonlinear components, the results seen in Fig. 1.15 are observed. In this case the fundamental output power will start first with a linear progression with the input power. That is, a 1-dB increase in input power will impose a 1-dB increase in output power. Next it will start compressing or expanding the growth slope accordingly to Eq. (1.59). The IMD power will start at a lower level than the fundamental power, since it depends on a third-order polynomial. Then it will rise at a slope of 3 dB for each additional 1 dB at the input, corresponding to the third-order polynomial. Finally, for higher values of output power, the IMD will compress or expand according to higher orders of distortion. If the linear output response on the one hand and the third-order small-signal response on the other hand are extrapolated, it gives rise to a FOM called the third-order intercept point ($\mathrm{IP}_3$). This FOM allows wireless engineers to calculate the small-signal nonlinear response very efficiently. Actually, this is very important for obtaining the amount of nonlinear distortion that arises from an interferer at the wireless system receiver.

DEFINITION 1.5 *The third-order intercept point ($IP_3$) is a fictitious point that is obtained when the extrapolated 1-dB/dB slope line of the output fundamental power intersects the extrapolated 3-dB/dB slope line of the IMD power.*

From a mathematical point of view, we will have to calculate the input or output power at which they intercept, thus, referring to Eq. (1.59),

$$\frac{1}{2}a_1^2 A_i^2 = \frac{9}{32}a_3^2 A_i^6 \rightarrow A_i^2 = \frac{4}{3}\frac{a_1}{a_3} \tag{1.65}$$

$$\mathrm{IP}_3 = P(\omega_1) = \frac{2}{3}\frac{a_1^3}{a_3} \tag{1.66}$$

In this case the output $\mathrm{IP}_3$ was calculated, but in certain cases it is preferable to calculate the input $\mathrm{IIP}_3$; see Fig. 1.15.

It should be mentioned that, despite their rarely being seen, some other intercept figures of merit could be defined for fifth-order ($\mathrm{IP}_5$) or seventh-order ($\mathrm{IP}_7$) distortion.

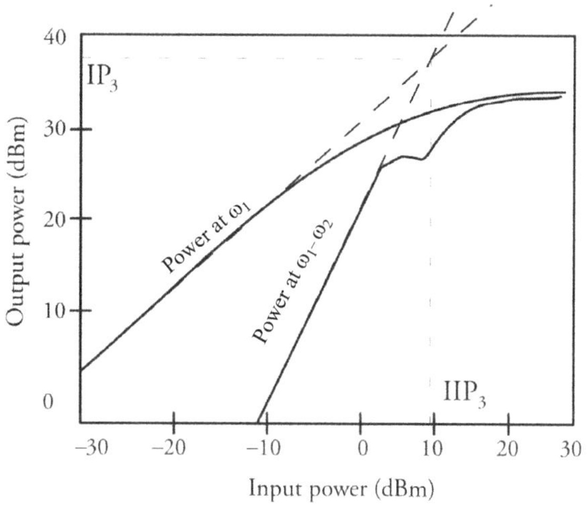


**Figure 1.15** The definition of IP3. The extrapolated IP3 can be seen as the intercept of the third-order and fundamental power rises.

$IP_3$ can be further used to calculate the intermodulation power at any input power, if restricted to the small-signal region. It is then possible to relate the IMR to $IP_3$ by [4]

$$\mathrm{IP3}_{\mathrm{dB}} = P_{\mathrm{fund}_{\mathrm{dB}}} + \frac{1}{2}\mathrm{IMR}_{\mathrm{dB}} \tag{1.67}$$

or

$$\mathrm{IMR}_{\mathrm{dB}} = 2\left(\mathrm{IP3}_{\mathrm{dB}} - P_{\mathrm{fund}_{\mathrm{dB}}}\right) \tag{1.68}$$

These equations were calculated for a two-tone input signal with equal amplitudes in both tones, and $P_{\mathrm{fund}_{\mathrm{dB}}}$ is the output power at a specific tone.

#### 1.5.2.4 Nonlinear distortion in the presence of dynamic effects

It should be stated here that certain nonlinear DUTs present what are called memory effects. These effects are a representation of dynamics in the nonlinear generation mechanism, as is discussed in [6]. The dynamic effects can mask the real intermodulation distortion in the DUT, since the lower sideband and higher sideband of the two-tone analysis can be different. Most often the dynamics arise from a baseband component being mixed with the fundamentals. So the dynamic effects can be measured by exciting these baseband components.

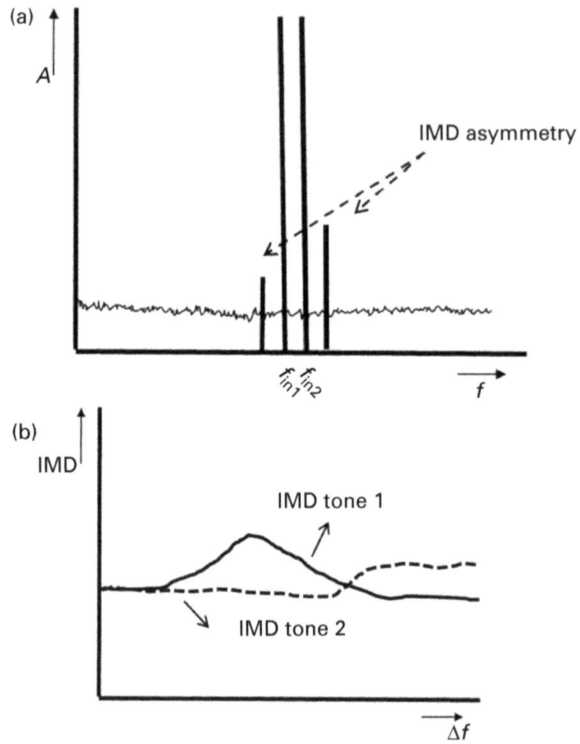


**Figure 1.16** The impact of two-tone nonlinear dynamics: (a) two-tone IMD measurement when memory effects are visible; and (b) two-tone IMD variation with tone spacing.

These phenomena can be measured in the laboratory by exciting the nonlinear device with a signal that covers most of the baseband spectrum behavior. One way to achieve that is by using a two-tone signal and varying the tone spacing between the tones. If the IMD is measured with varying tone separation, then the impact of the baseband envelope behavior will be seen in the IMD variation, as shown in Fig. 1.16.

### 1.5.3 FOMs for nonlinear continuous spectra

New wireless communication standards, mainly digital wireless communications, have great richness in spectral content, which means that the single-tone and two-tone figures of merit have become obsolete, and are not able to characterize important aspects of that type of transmission.

Thus microwave and wireless system engineers have started to use other forms of excitation and other types of FOM to account for nonlinear distortion in digital wireless communication signals. Moreover, the nonlinear nature of RF components means that there will be a close relationship between the usefulness of a certain characterization technique and the similarity of the test signal to the real equipment's excitation.

In this sense engineers are considering other forms of excitation, including digitally modulated carriers with pseudo-random baseband signals, multi-tones (more than two

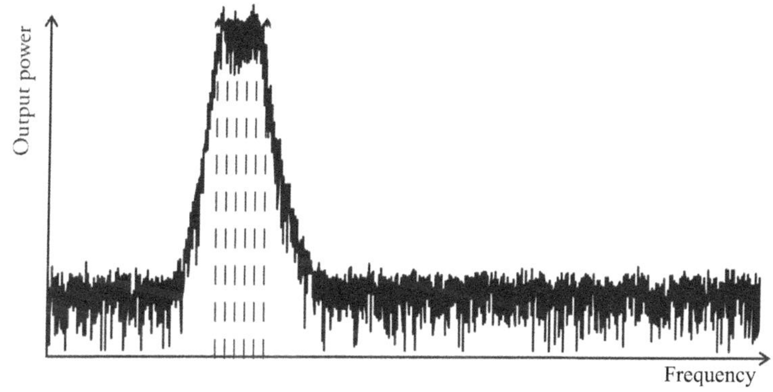


**Figure 1.17** A typical input signal in a wireless communication system.

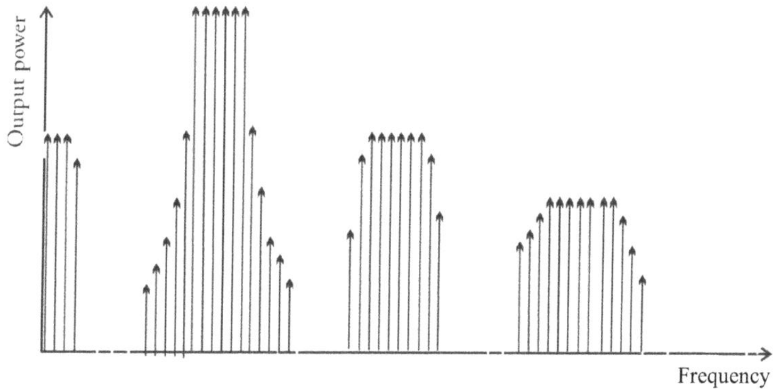


**Figure 1.18** The nonlinear output response of a multi-sine signal excitation.

tones, usually called multi-sines), and band-limited noise [7]. In this section we will address figures of merit developed for rich spectra, continuous or not.

In Fig. 1.17 we can see the typical input in a wireless communication system, for which the spectrum is actually continuous. Microwave engineers sometimes use similar signals in the laboratory for mimicking this type of spectral richness using a multi-sine signal, Fig. 1.18, which allows a much simpler analysis of the output signal.

Figures 1.18 and 1.19 present the output of a typical signal like this, where the baseband, IMD, second harmonic and third harmonic are evident. This is similar to what we had previously seen in a two-tone excitation, but now each cluster is much richer.

In this sense the out-of-band distortion is dealt with by deploying the same line of thought as that which was used for the single-tone and the two-tone signal, meaning that in traditional wireless signals these are eliminated using output filtering. So the typical FOM for wireless systems usually refers mainly to the in-band components.

Figure 1.20 presents the in-band distortion that can be seen at the output of a nonlinear system when it is excited by a bandlimited continuous spectrum.

It is evident that we can identify the co-channel distortion that falls on top of the linear output signal, corresponding to a linear complex gain multiplication of the input

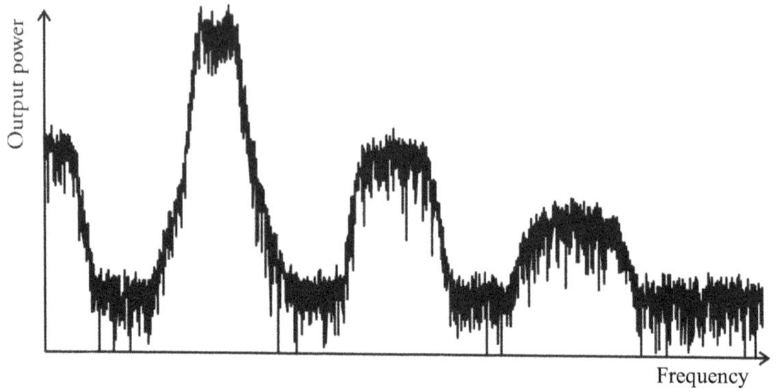


**Figure 1.19** The nonlinear output response of a rich input spectrum.

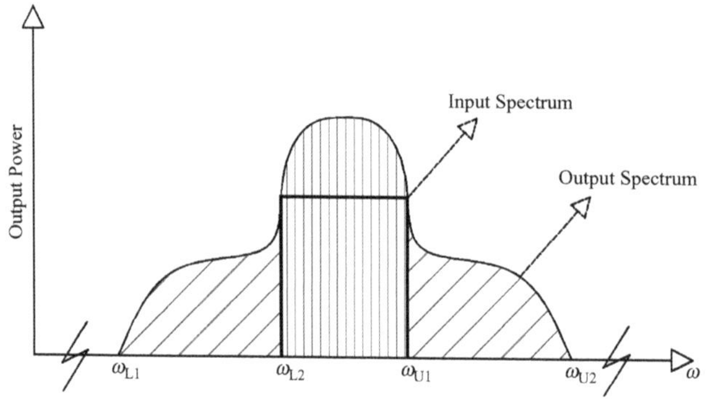


**Figure 1.20** The in-band nonlinear output response of a rich input spectrum.

spectrum. We can also see a sideband on each side, corresponding to what is called spectral regrowth. It arises from the nonlinear odd-order terms, similarly to the IMD tones appearing in a two-tone excitation. Having this signal in mind, we can now define some important figures of merit for characterization of nonlinear rich spectra.

#### 1.5.3.1 The multi-sine intermodulation ratio

The multi-sine intermodulation ratio (M-IMR) is actually a generalization of the IMR concept introduced in Section 1.5.2.1. As can be seen from Fig. 1.21, this figure of merit can be defined as follows.

DEFINITION 1.6 *The multi-sine intermodulation ratio (M-IMR) is defined as the ratio of the common fundamental power per tone, $P_{\text{fund}_{\text{tone}}}$, to the power of the $\omega_r$ distortion component present in the lower or upper adjacent bands, $P_{\text{L/U}}(\omega_{\text{r}})$.*

In mathematical terms it is nothing other than

$$\text{M-IMR} = \frac{P_{\text{fund}_{\text{tone}}}}{P_{\text{L/U}}(\omega_{\text{r}})} \tag{1.69}$$

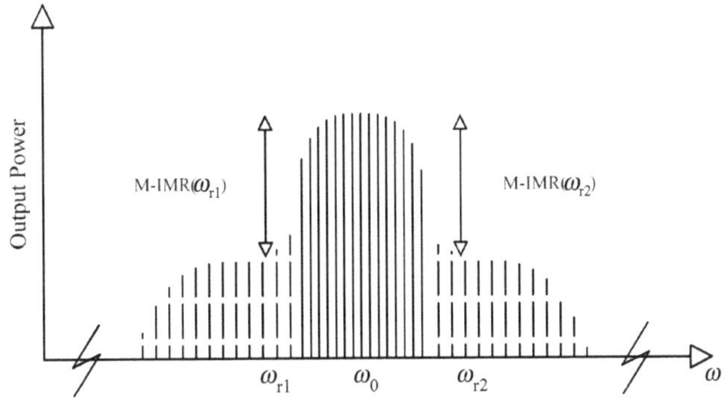


**Figure 1.21** The definition of the multi-sine intermodulation ratio.

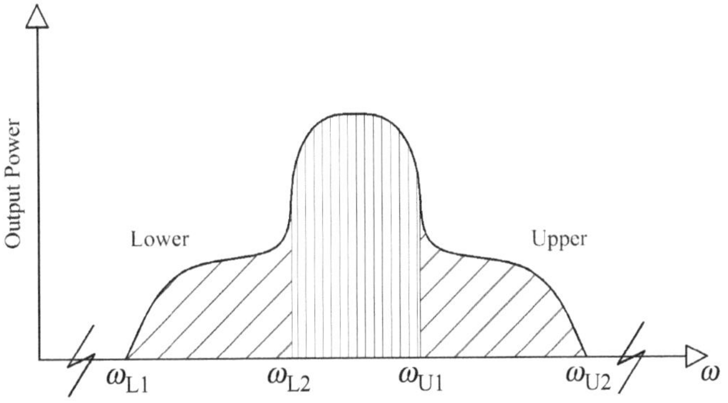


**Figure 1.22** The definition of the adjacent-channel power ratio.

### 1.5.3.2 The adjacent-channel power ratio

The FOM known as the M-IMR allows engineers to measure and account for each tone in the multi-sine approach, but when the signal is continuous the user should account not for each sine, but for all the spectral power that is being created by the nonlinearity. This part of the spectrum is called adjacent-channel distortion, and is composed of all distortion components falling on the adjacent-channel location. Actually, in typical communication scenarios it can be a source of interference with adjacent channels. For accounting for this type of distortion, and mainly the amount of power being regrown, several FOMs can be defined.

DEFINITION 1.7 *The total adjacent-channel power ratio ($ACPR_T$) is the ratio of the total output power measured in the fundamental zone, $P_{\mathrm{fund}}$, to the total power integrated in the lower, $P_{\mathrm{LA}}$, and upper, $P_{\mathrm{UA}}$, adjacent-channel bands.*

Figure 1.22 shows this FOM and how it is calculated. Mathematically it can be described as

$$\mathrm{ACPR_T} = \frac{P_{\mathrm{fund}}}{P_{\mathrm{LA}} + P_{\mathrm{UA}}} = \frac{\int_{\omega_{L_2}}^{\omega_{U_1}} S_o(\omega)d\omega}{\int_{\omega_{L_1}}^{\omega_{L_2}} S_o(\omega)d\omega + \int_{\omega_{U_1}}^{\omega_{U_2}} S_o(\omega)d\omega} \tag{1.70}$$

where $S_o(\omega)$ is the power spectral density.

Sometimes it is also interesting to address only a specific part of the spectral regrowth, and in that situation we can define the upper or lower ACPR value. This FOM can be defined as follows.

DEFINITION 1.8 *The upper or lower adjacent-channel power ratio ($ACPR_L$ or $ACPR_U$) is the ratio between the total output power measured in the fundamental zone, $P_{\mathrm{fund}}$, and the lower or upper adjacent-channel power, $P_{\mathrm{LA}}$ or $P_{\mathrm{UA}}$.*

Mathematically,

$$\begin{aligned} \mathrm{ACPR_L} &= \frac{P_{\mathrm{fund}}}{P_{\mathrm{LA}}} = \frac{\int_{\omega_{L_2}}^{\omega_{U_1}} S_o(\omega)d\omega}{\int_{\omega_{L_1}}^{\omega_{L_2}} S_o(\omega)d\omega} \\ \mathrm{ACPR_U} &= \frac{P_{\mathrm{fund}}}{P_{\mathrm{UA}}} = \frac{\int_{\omega_{L_2}}^{\omega_{U_1}} S_o(\omega)d\omega}{\int_{\omega_{U_1}}^{\omega_{U_2}} S_o(\omega)d\omega} \end{aligned} \tag{1.71}$$

Sometimes it is preferable to consider not all of the adjacent-channel power, but only a piece of it. That happens because in continuous spectra it is usually difficult to define where the adjacent-channel spectrum starts and ends, mainly due to the roll-off of the system filters. Thus the industry refers to this FOM as the spot ACPR, this being defined as follows.

DEFINITION 1.9 *The spot adjacent-channel power ratio ($ACPR_{SP_L}$ or $ACPR_{SP_U}$) is the ratio of the total output power measured in the fundamental zone, $P_{\mathrm{fund}}$, to the power integrated in a band of predefined bandwidth and distance from the center frequency of operation $P_{\mathrm{SP}_{L/U}}$.*

Mathematically it can be described as

$$\begin{aligned} \mathrm{ACPR_{SP_L}} &= \frac{P_{\mathrm{fund}}}{P_{\mathrm{SP_L}}} = \frac{\int_{\omega_{L_2}}^{\omega_{U_1}} S_o(\omega)d\omega}{\int_{\omega_{\mathrm{NBL}_1}}^{\omega_{\mathrm{NBL}_2}} S_o(\omega)d\omega} \\ \mathrm{ACPR_{SP_U}} &= \frac{P_{\mathrm{fund}}}{P_{\mathrm{SP_U}}} = \frac{\int_{\omega_{L_2}}^{\omega_{U_1}} S_o(\omega)d\omega}{\int_{\omega_{\mathrm{NBU}_1}}^{\omega_{\mathrm{NBU}_2}} S_o(\omega)d\omega} \end{aligned} \tag{1.72}$$

Figure 1.23 shows this figure of merit and how it is calculated.

#### 1.5.3.3 Co-channel distortion FOMs

As was seen previously, nonlinear distortion generates adjacent-band spectral regrowth, but also co-channel distortion, which imposes a strong degradation of the signal-to-noise ratio. Unfortunately, co-channel distortion falls exactly on top of the input signal spectrum, and thus on top of the linear output signal.

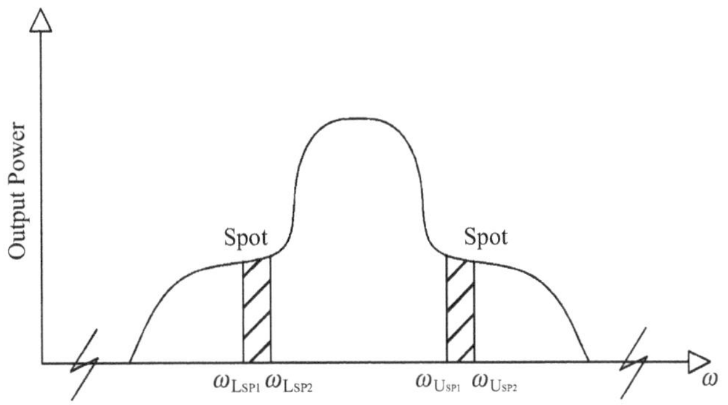


**Figure 1.23** The definition of the adjacent-channel spot power ratio.

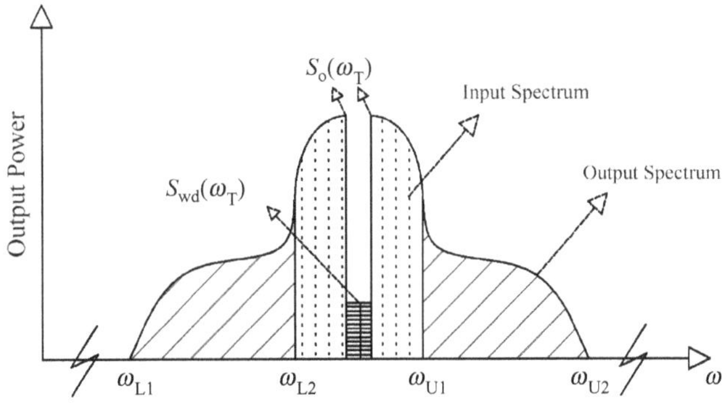


**Figure 1.24** The definition of the noise power ratio.

This type of distortion can be accounted for using a FOM known as the co-channel power ratio (CCPR), which allows a correct measure of the nonlinear distortion falling inside the band. However, because this form of distortion is intricately mixed with the fundamental zone of much higher amplitude, the measurement of this type of distortion is quite difficult. That is why wireless system engineers came up with a FOM called the noise power ratio (NPR) that actually allows one to characterize the co-channel distortion in an indirect way. The basic idea is mainly to open a notch in the input signal spectrum, and to account for the noise in that notch hole both at the input and at the output, Fig. 1.24. We will see in future chapters how to measure this type of co-channel distortion, but let us now define these co-channel FOMs.

The NPR can be defined as follows.

DEFINITION 1.10 *The noise power ratio (NPR) is defined as the ratio of the output power spectral density function measured in the vicinity of the test window position,* $\omega_T$, $S_o(\omega_T)$, *to the power spectral density observed within that window,* $S_{wd}(\omega_T)$.

Mathematically it is expressed by

$$\mathrm{NPR}(\omega_T) = \frac{S_o(\omega_T)}{S_{wd}(\omega_T)} \tag{1.73}$$

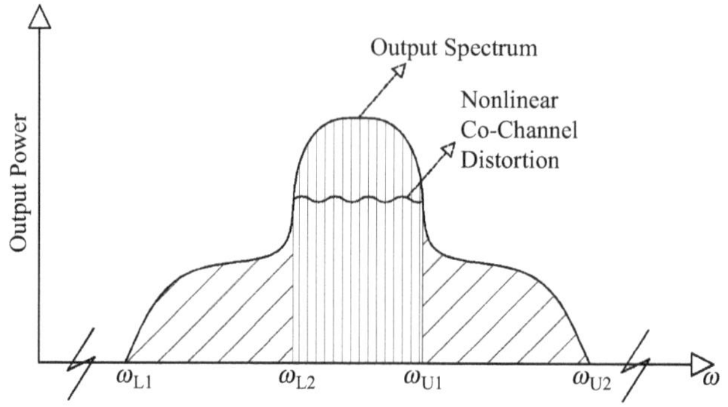


**Figure 1.25** The definition of the co-channel power ratio.

The co-channel power ratio (Fig. 1.25) can be defined as follows.

DEFINITION 1.11 *The co-channel power ratio (CCPR) is defined as the ratio of the integrated output power measured in the fundamental zone, $P_{\text{fund}}$, to the total integrated co-channel perturbation, $P_{\text{co-channel}}$.*

Mathematically it is expressed as

$$\text{CCPR} = \frac{P_{\text{fund}}}{P_{\text{co-channel}}} = \frac{\int_{\omega_{L_2}}^{\omega_{U_1}} S_o(\omega)d\omega}{\int_{\omega_{L_2}}^{\omega_{U_1}} S_{\text{co-channel distortion}}(\omega)d\omega} \tag{1.74}$$

## 1.6 System-level FOMs

Regarding general figures of merit that can be applied to generic RF and wireless components and circuits, some system-level FOMs will now be presented. These FOMs relate more generally not to a single component or circuit, but rather to a bigger system. In that sense they are normally stated as high-level quantities, and most of the time they are related to the information being sent over the communication channel and thus not necessarily to any spectral or time characteristics.

Some of these FOMs include the error-vector magnitude and the bit error rate, which are FOMs that can be measured at a high level of abstraction and that depend not on a single component but rather on the activity of the complete system.

### 1.6.1 The constellation diagram

In a digital radio, the evaluation of the transmitted signals is fundamental. This characterization can be done by referring to the constellation diagram. Let us explain the concept of a constellation diagram. In a digital modulated RF signal we can describe the input and output signal as a sine wave that is in phase or in quadrature phase arrangement:

$$\begin{aligned} x(t) &= A(t)\cos(\omega t + \theta(t)) \\ &= A(t)\frac{e^{j(\omega t+\theta(t))} + e^{-j(\omega t+\theta(t))}}{2} \\ &= \mathrm{Re}\left[A(t)e^{-j(\omega t+\theta(t))}\right] \\ &= \mathrm{Re}\left[A(t)e^{-j\theta(t)}e^{-j\omega t}\right] \\ &= \mathrm{Re}\left[\tilde{x}(t)e^{-j\omega t}\right] \\ &= \mathrm{Re}\left\{[I(t) + jQ(t)]e^{-j\omega t}\right\} \\ &= I(t)\cos(\omega t) + Q(t)\sin(\omega t) \end{aligned} \tag{1.75}$$

By representing a wireless signal as a complex number and modulating a cosine and sine carrier signal with the real and imaginary parts, respectively, the symbol can be sent with orthogonal carriers on the same frequency.

These carriers are often referred to as quadrature carriers. If the wireless system uses a coherent detector, then it is possible to independently demodulate these carriers. This principle is actually the base for quadrature modulation.

Actually the phase and quadrature information is normally called the complex envelope of the signal, and is represented in several ways. One possibility is to include a two-time-domain graph with $I(t)$ and $Q(t)$ plotted over time; the other possibility is using a constellation diagram, where the phase and quadrature values are plotted over each other in a complex graph representation (see Fig. 1.26).

Since the symbols are represented as complex numbers, they can be visualized as points on the complex plane. The real-number and imaginary-number coordinate axes are often called the in-phase axis, or $I$-axis, and the quadrature axis, or $Q$-axis.

Plotting several symbols in a scatter diagram produces the constellation diagram. The points on a constellation diagram are called constellation points. They are a set of modulation symbols comprising the modulation alphabet.

Thus a constellation diagram is nothing more than a representation of these $I(t)$ and $Q(t)$ in a complex diagram, plotting each pattern in phase ($I(t)$) and quadrature ($Q(t)$). Actually, it displays the signal as a two-dimensional scatter diagram in the complex plane. Sometimes the plot is displayed only at the symbol-sampling instants, but the overall trajectory is also very important for understanding certain aspects of the system behavior.

In a pure quadrature phase-shift keying (QPSK) signal the constellation diagram will be that presented in Fig. 1.27(a) if the represented symbols are plotted only at the symbol rate, but it will be that presented in Fig. 1.27(b) if the overall trajectories of $I(t)$ and $Q(t)$ are plotted.

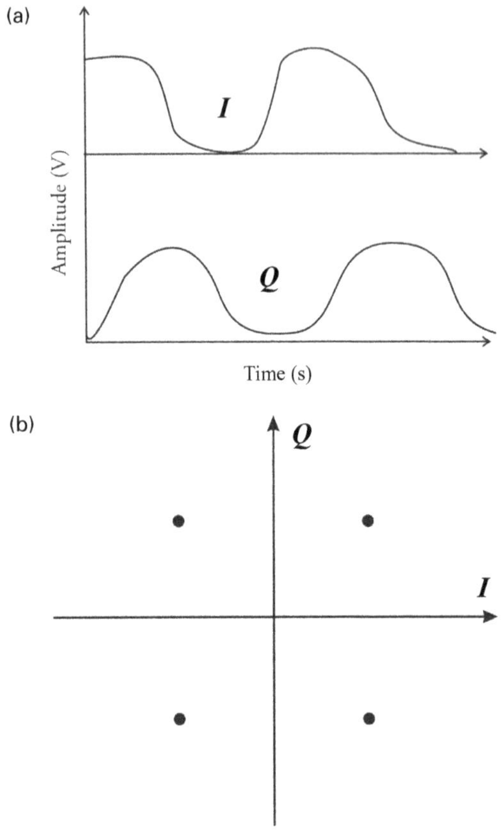


**Figure 1.26** The $I(t)$ and $Q(t)$ representation: (a) time-domain waveforms and (b) the constellation diagram.

Constellation diagrams can be used to recognize the type of interference and distortion in a signal.

This tool is very important, since it allows engineers to observe the transmitted constellation points and to compare them with the received ones. In this way, they are able to identify the similarities and differences between them, thereby accounting for the signal degradation from a point of view that is actually the ultimate one, meaning that it relates directly to the information transmission quality.

The system itself can degrade the transmitted signal by adding noise to the signal, or can degrade it due to nonlinear distortion, as was seen in Section 1.4.

In terms of transmission degradation, engineers will seek in the constellation diagram a deviation of the actual received signal from the transmitted one, and will calculate this difference using some form of Euclidean distance. Thus the receiver will demodulate the received signal incorrectly if the corruption has caused the received symbol to move closer to another constellation point than the one transmitted.

This is actually called maximum-likelihood detection. The use of the constellation diagram allows a straightforward visualization of this process.

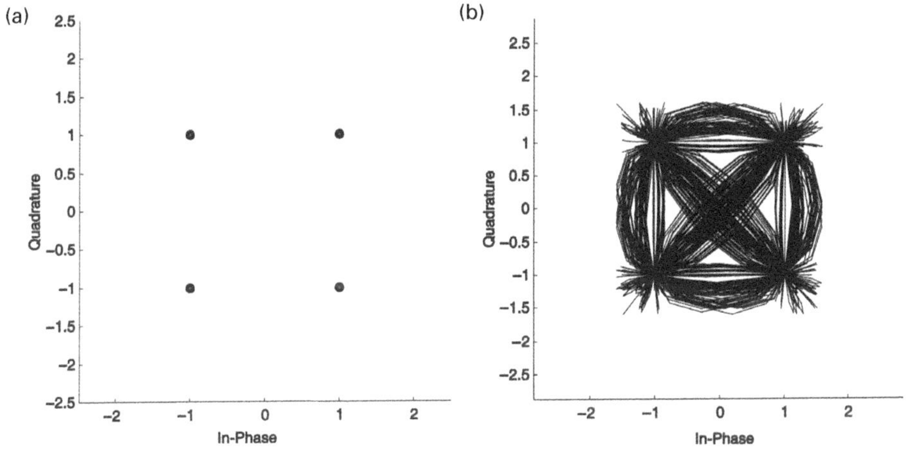


**Figure 1.27** The QPSK constellation diagram: (a) Sampled at the symbol rate; and (b) the overall trajectory.

## 1.6.2 The error-vector magnitude

Considering the constellation-diagram approach, the first system-level FOM to be presented is the error-vector magnitude (EVM). The EVM is a measure that is used to evaluate the performance of an RF system in digitally modulated radios.

An ideal signal sent by an ideal transmitter without any interference will have all constellation points precisely at the ideal locations. However, if the signal is interfered with by different propagation-channel imperfections, such as noise, nonlinear distortion, phase noise, adjacent-channel interference, etc., the symbols and thus the constellation points will deviate from the ideal locations.

The EVM is actually accounting for the errors in the points in a constellation diagram. It is nothing more than a measure of how far the points are from their ideal locations.

DEFINITION 1.12 *The error-vector magnitude (EVM) is a vector (geometric) in the I–Q plane between the ideal constellation point and the point received by the receiver. It can also be stated as the difference between actually received symbols and ideal symbols. The average power of the error vector, normalized with respect to the signal power, is the EVM. For the percentage format, the root-mean-square* (rms) *average is used.*

In mathematical terms the EVM is expressed as a percentage:

$$\mathrm{EVM}_{[\mathrm{RMS}]} = \sqrt{\frac{\sum_{k=1}^{N} |Z_{\mathrm{c}}(k) - S(k)|^2}{\sum_{k=1}^{N} |S(k)|^2}} \tag{1.76}$$

$$\mathrm{EVM}_{[\%]} = \mathrm{EVM}_{[\mathrm{RMS}]} \times 100 \tag{1.77}$$

where $N$ is the number of received symbols, $Z_{\mathrm{c}}(\ )$ is the actual received symbol, and $S(\ )$ is the ideal symbol that should be received (Fig. 1.28).

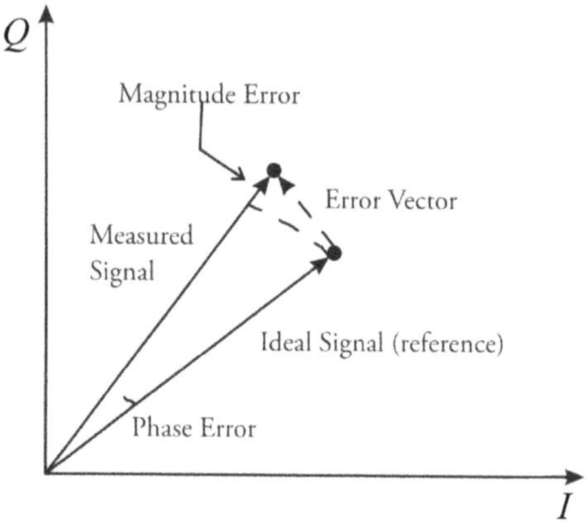


**Figure 1.28** Calculation of the error-vector magnitude.

The EVM can also be obtained from the signal-to-noise ratio, and some relationships have been developed to compare these two quantities [8]:

$$\mathrm{EVM}_{[\mathrm{RMS}]} = \sqrt{\frac{1}{\mathrm{SNR}}} \tag{1.78}$$

This equation is very important, since it states that, by accounting for the noise degradation or the nonlinear distortion degradation inside the band, it is possible to account for the EVM of an overall system.

### 1.6.3 The peak-to-average power ratio

Another important FOM that actually cannot be attributed to the system itself, but rather must be attributed to the signal, is the one that accounts for the relationship between the peak power and the average power of the signal. Some authors [9] have used the peak-to-average power ratio (PAPR), as the ratio of the average power that would result if the envelope were sustained at its peak magnitude to the average power in the $N$-sinusoid sum. The PAPR thus has the mathematical form

$$\mathrm{PAPR} = \frac{\max|x(t)^2|}{[1/(NT)] \int_0^{NT} |x(t)|^2\, dt} \tag{1.79}$$

A version of the PAPR for a sampled signal can also be used. This is defined as

$$\mathrm{PAPR_s} = \frac{\max|x_k^2|}{E|(x(k)|^2\, dt} \tag{1.80}$$

where $E(\,)$ is the expectancy of $x$.

## 1.7 Filters

Filters, combiners, power dividers, isolators, and other linear components can be characterized using the linear FOMs presented in Section 1.3. In all these situations the

**Table 1.2** A filter datasheet of electrical characteristics (guaranteed over $-50\,^{\circ}\mathrm{C}$ to $+90\,^{\circ}\mathrm{C}$ operating temperature)

| Part number | Frequency band [MHz] | Insertion loss (dB) | Maximum VSWR (dB) | Typical attenuation (dB) |
|---|---|---|---|---|
| Filter 1 | 800–1000 | 0.35 typical (0.5 max.) | 1.5 | 30 at $2F_0$ |
| Filter 2 | 865–985 | 0.34 typical (0.5 max.) | 1.4 | 27 at $2F_0$ |
| Filter 3 | 1700–1900 | 0.37 typical (0.5 max.) | 1.6 | 40 at $2F_0$ |

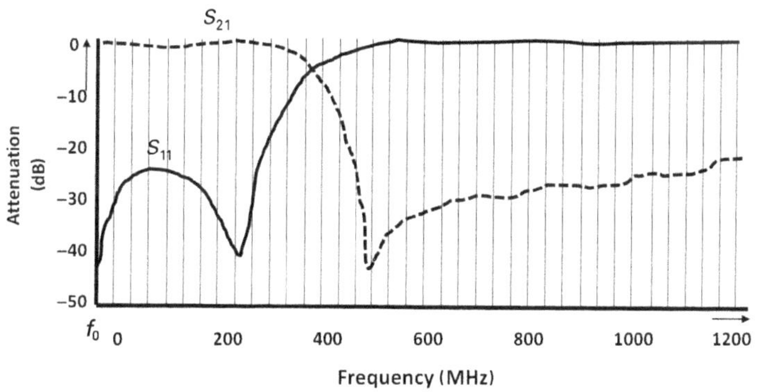


**Figure 1.29** Measured $S$-parameters of a low-pass filter.

main FOMs to consider are the $S$-parameters, and correspondingly the insertion loss and VSWR as well as the bandwidth.

In order to better understand a typical datasheet of a filter, consider Table 1.2. On this example datasheet, several filters, all bandpass filters, are presented, each with their respective bandwidth, insertion loss, in this case near 0.5 dB in the band of interest, and VSWRs on the order of 1.4, which corresponds to an impedance mismatch of near $\Gamma_{\mathrm{in}} = 0.17$ and to an impedance of 35 Ω to 70 Ω in a 50-Ω environment.

The bandwidth of a filter is usually determined by the lower and upper frequencies at which the in-band insertion loss has dropped with a certain dB value, typically 3 dB.

Another important FOM is the one corresponding to the out-of-band attenuation. In this sample case, it is specified at the double frequency of $F_0$, which is the bandwidth's upper frequency limit. These frequencies are indicated in Fig. 1.29, showing the measured $S$-parameters of a low-pass filter.

## 1.8 Amplifiers

Amplifiers have been the main components in any RF/microwave radio design, since they allow one to increase the signal level in order to be able to transmit or receive the communication signal with a certain signal-to-noise ratio value.

Depending on the section of the system where the amplifier has to be inserted, the amplifier can have different functions, for instance consider Fig. 1.1. If the amplifier is inserted into the receiver as its first stage, then the focus will be on the low-noise behavior, and therefore it is a low-noise amplifier (LNA). On the other hand, if the amplifier is to be inserted into the transmitter as its last stage, the amplifier is focused on increasing the output power, since the main objective of the transmitter is to achieve a high power level in order to fulfill the link budget that the system engineer has designed. In this case the amplifier is called a power amplifier (PA). In other sections of the transceiver chain, we can see some generic amplifiers (GAs) that are usually designed to maximize gain, and not necessarily for high power or low noise. This category of amplifier includes the variable-gain amplifiers (VGAs). Their objective is to reduce distortion in subsequent stages, by dynamically optimizing the gain in order to maintain a constant output amplitude.

In the next sections, the FOMs applicable to amplifiers are described. A distinction is made among linear and noise FOMs, which are of interest primarily for GAs and LNAs but also for PAs; nonlinear FOMs, which are primarily applicable to PAs; and transient FOMs, which are specific for VGAs. As an example, Table 1.3 presents a typical datasheet of an amplifier. The linear FOMs calculable from $S$-parameters are the gain as function of frequency, gain variation over temperature, input and output return loss, and reverse isolation. The noise figure is listed as well. The nonlinear FOMs listed in this example datasheet are the 1-dB-compression point, saturated power, and third-order intercept point. An amplifier datasheet always includes also the DC operating conditions. The temperature dependency is especially important for PAs.

## 1.8.1 Linear and noise FOMs

Since amplifiers are two-port networks, the linear FOMs described in Section 1.3 are generally applicable.

First of all, an RF engineer should be aware of any mismatch in the input and output connections, in order to guarantee that the power loss in these connections is minimal. This can be expressed by the VSWR, defined in Section 1.3.1.1, or by the return loss, RL, defined in Section 1.3.1.2.

The FOM related to noise is the noise figure, NF, which was defined in Section 1.3.2.1. The lower the noise figure, the better the performance of the low-noise amplifier. Nevertheless, even when not optimized, the noise figure is usually listed as well in the datasheets of the other types of amplifiers. It is a measure for the noise added by the amplifier in the system, which is of importance to evaluate the overall noise, as expressed by the Friis formula (1.31).

The aim of an amplifier is to amplify the input signal, so the gain is its most characteristic FOM. Before elaborating on the gain, it is also important to note that an amplifier should act as an isolator, or at least as a strong attenuator, in the reverse direction. Or, in other words, an excitation at its output, e.g., caused by a malfunctioning subsequent block in the chain, should not propagate to its input because this may damage the

**Table 1.3** Amplifier datasheet, electrical characteristics (specified at 25 °C and 50 mA)

| Parameter | | Minimum | Typical | Maximum | Units |
|---|---|---|---|---|---|
| Frequency range | | DC | | 10 | GHz |
| Gain | $f = 0.1$ GHz | 11.5 | 12.5 | 13 | dB |
| | $f = 1$ GHz | | 12.3 | | |
| | $f = 2$ GHz | 10 | 11.5 | 12.9 | |
| | $f = 6$ GHz | | 11.1 | | |
| | $f = 10$ GHz | | 10.8 | | |
| Gain versus temperature | $f = 0.1$ GHz | | 0.001 | 0.002 | dB/°C |
| | $f = 1$ GHz | | 0.001 | 0.003 | |
| | $f = 2$ GHz | | 0.0015 | 0.0035 | |
| | $f = 6$ GHz | | 0.0019 | 0.0038 | |
| | $f = 10$ GHz | | 0.0022 | 0.004 | |
| Input return loss | $f = 0.1$ GHz | | 30 | | dB |
| | $f = 1$ GHz | | 25 | | |
| | $f = 2$ GHz | | 20 | | |
| | $f = 6$ GHz | | 19 | | |
| | $f = 10$ GHz | | 17 | | |
| Output return loss | $f = 0.1$ GHz | | 26 | | dB |
| | $f = 1$ GHz | | 23 | | |
| | $f = 2$ GHz | | 21 | | |
| | $f = 6$ GHz | | 16 | | |
| | $f = 10$ GHz | | 15 | | |
| Reverse isolation | $f = 2$ GHz | 12 | 18 | | dB |
| Output 1-dB-compression point | $f = 0.1$ GHz | | 15 | | dBm |
| | $f = 1$ GHz | | 15 | | |
| | $f = 2$ GHz | | 15 | | |
| | $f = 6$ GHz | | 15 | | |
| | $f = 10$ GHz | | 11 | | |
| Saturated output power (at 3 dB compression) | $f = 0.1$ GHz | | 16 | | dBm |
| | $f = 1$ GHz | | 16 | | |
| | $f = 2$ GHz | | 16 | | |
| | $f = 6$ GHz | | 15 | | |
| | $f = 10$ GHz | | 14 | | |
| Output $IP_3$ | $f = 0.1$ GHz | 24 | 30 | | dBm |
| | $f = 1$ GHz | 24 | 30 | | |
| | $f = 2$ GHz | 24 | 30 | | |
| | $f = 6$ GHz | 24 | 29 | | |
| | $f = 10$ GHz | 23 | 27 | | |
| Noise figure | $f = 0.1$ GHz | | 4 | 5 | dB |
| | $f = 1$ GHz | | 4.2 | 5 | |
| | $f = 2$ GHz | | 4.2 | 5 | |
| | $f = 6$ GHz | | 4.4 | 5.2 | |
| | $f = 10$ GHz | | 4.6 | 5.5 | |
| Group delay | $f = 2$ GHz | | 60 | | ps |
| Operating current | | | 50 | | mA |
| Operating voltage | | 3 | 3.3 | 4 | V |
| Voltage variation with temperature | | | −2 | | mV/°C |
| Voltage variation with current | | | 9 | | mV/mA |
| Thermal resistance | | | 180 | | °C/W |

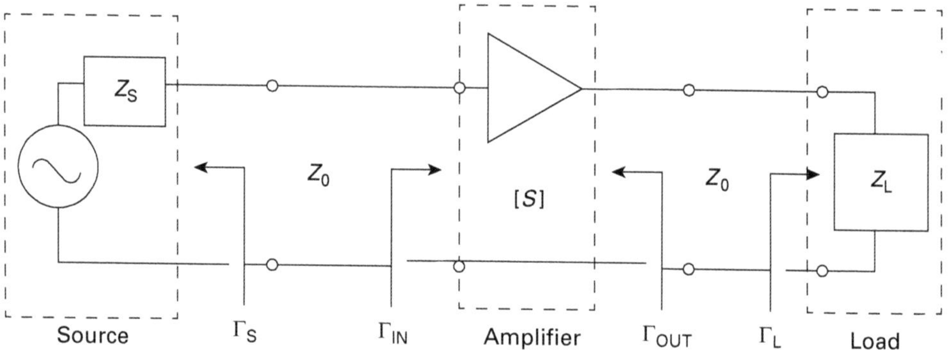


**Figure 1.30** Power gain definitions.

preceding blocks. The corresponding FOM is the isolation. It is related to $S_{12}$ since this is the reverse transmission coefficient (see Section 1.2):

$$\text{isolation} = \left|20 \log_{10}|S_{12}|\right| \text{ dB} \tag{1.81}$$

So intuitively one may think that the gain of an amplifier is expressed by its $S_{21}$. It is a more complex matter, though. In fact, there are three power gain definitions, which are the transducer power gain $G_T$, the operating power gain $G$, and the available power gain $G_A$. The three definitions are applicable to any amplifier, but a given type of power gain may be more suitable for particular cases, as we will describe below [1, 2].

Before introducing the definitions of these power gains, we first should introduce some other definitions, explained by means of Fig. 1.30.

Figure 1.30 shows a microwave amplifier that is excited by a source with source impedance $Z_S$ and terminated with a load $Z_L$. In a system design, the load represents the input impedance of the circuit block following the amplifier, such as the antenna in a transmitter. Similarly, $Z_S$ may represent the output impedance of the circuit block preceding the amplifier. In the subsequent equations, the amplifier is represented by its $S$-parameters. The source impedance $Z_S$ corresponds to the reflection coefficient $\Gamma_S$, and similarly the load $Z_L$ corresponds to the reflection coefficient $\Gamma_L$.

Two other reflection coefficients are indicated in Fig. 1.30, namely $\Gamma_{IN}$ and $\Gamma_{OUT}$. The reflection coefficient $\Gamma_{IN}$ is the input reflection coefficient of the amplifier followed by the load. Similarly, the reflection coefficient $\Gamma_{OUT}$ is the output reflection coefficient of the amplifier preceded by the source impedance. $\Gamma_{IN}$ and $\Gamma_{OUT}$ can be expressed in terms of the blocks' parameters, as follows (see also Eq. (1.10)):

$$\Gamma_{IN} = S_{11} + \frac{S_{12}S_{21}\Gamma_L}{1 - S_{22}\Gamma_L} \tag{1.82}$$

$$\Gamma_{OUT} = S_{22} + \frac{S_{12}S_{21}\Gamma_S}{1 - S_{11}\Gamma_S}$$

Note that $\Gamma_{IN}$ becomes independent of the load if the amplifier is unilateral. This means that there is no feedback from output to input, which corresponds to $S_{12}$ equal to zero. The measure is the isolation FOM (Eq. (1.81)). Similarly, $\Gamma_{OUT}$ is independent of the reflection coefficient of the source network, $\Gamma_S$, if the amplifier is unilateral. Also, $\Gamma_{IN}$ reduces to $S_{11}$ if $\Gamma_L$ is equal to zero, or in other words $Z_L$ is equal to 50 Ω. Similarly, $\Gamma_{OUT}$ reduces to $S_{22}$ if $\Gamma_S$ is equal to zero, or, in other words, the source impedance is 50 Ω.

Before proceeding to the power gain definitions, we still need to define several powers. The power available from the source is denoted as $P_{AVS}$. The power effectively going into the amplifier is $P_{IN}$. $P_{IN}$ is lower than $P_{AVS}$ if there is a mismatch between the source and the input of the amplifier. If there is no mismatch, $P_{IN}$ is equal to $P_{AVS}$. Similarly, at the output $P_{AVN}$ stands for the output power available from the amplifier, while $P_L$ is the power delivered to the load. $P_L$ is smaller than $P_{AVN}$ unless there is no mismatch between the amplifier output and the load.

The first power gain definition is the transducer power gain ($G_T$).

DEFINITION 1.13 *The transducer power gain ($G_T$) is the ratio of the power delivered to the load to the power available at the source.*

$G_T$ is expressed by the following equation:

$$G_T = \frac{P_L}{P_{AVS}}$$

$$= \frac{(1-|\Gamma_S|)^2}{|1-\Gamma_S\Gamma_{IN}|^2}|S_{21}|^2\frac{(1-|\Gamma_L|)^2}{|1-S_{22}\Gamma_L|^2} \tag{1.83}$$

$$= G_S G_0 G_L \tag{1.84}$$

The expression has been written as a product of three factors in order to be able to better evaluate the contribution of each block. The first factor, $G_S$, relates to the interaction between the source network and the input of the amplifier, $G_0$ stands for the contribution of the amplifier itself, and $G_L$ is related to the interaction between the output of the amplifier and the load. If the source and load are perfect, meaning $Z_S = Z_L = 50$ Ω, or $\Gamma_S = \Gamma_L = 0$, then $G_T$ reduces to $|S_{21}|^2$. In other words, the power gain definitions take into account the mismatches between the amplifier and its preceding and following blocks. In the case of $G_T$, the mismatches both at the input and at the output are taken into account. The transducer power gain is the best applicable in the real situation when one wants to know how much power effectively gets delivered to the load when the source generates a certain power level. As we will see next, the operating power gain stresses the mismatch at the output, while the available power gain relates to the mismatch at the input. So the aim of the various power gain definitions is to express the actual power gain when the amplifier is embedded in a system. In the limit when there are no mismatches at input and output, the three power gains reach their maximal values and are equal to each other, $G_{T,max} = G_{A,max} = G_{max}$.

The next definition is the operating power gain $G$.

DEFINITION 1.14 *The operating power gain (G) is the ratio of the power delivered to the load to the power going into the amplifier.*

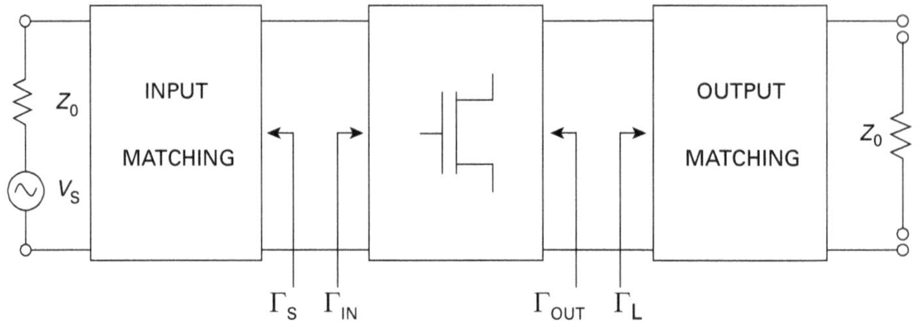


**Figure 1.31** The general scheme for amplifier design.

The corresponding expression is

$$G = \frac{P_L}{P_{IN}} = \frac{1}{1-|\Gamma_{IN}|^2}|S_{21}|^2\frac{1-|\Gamma_L|^2}{|1-S_{22}\Gamma_L|^2} \tag{1.85}$$

We note that $G$ is a function of $\Gamma_L$, while it is independent of $\Gamma_S$. In systems, $Z_{load}$ is usually close to 50 Ω, and therefore the differences between the various power gain values are small.

However, the definitions of power gain are generally applicable to two-port networks, and are therefore also made use of during amplifier design. In such cases, Fig. 1.31 is appropriate. The amplifier is now represented as a cascade of three blocks: a central block that includes the active part, namely one or more transistors enabling the amplification, while the other two blocks are the passive input and output matching networks. Note that an actual amplifier design may be more complicated than this general scheme. Here again, the central block is represented by its $S$-parameters. The input matching network is represented by its reflection coefficient $\Gamma_S$, and similarly the output matching network is represented by its reflection coefficient $\Gamma_L$. It is assumed in the calculations that the source has no mismatch, or, in other words, that its impedance is $Z_0$. The same applies to the output, where it is assumed that the terminating load is a $Z_0$ impedance. So the operating power gain is typically of interest for power amplifiers, because in such designs the load $Z_L$ is optimized for high power. Usually this value is not matched to the output impedance of the central active block, and therefore it reduces the operating power gain.

Finally, the available power gain $G_A$ is defined as follows.

DEFINITION 1.15 *The available power gain ($G_A$) is the ratio of the power available from the amplifier to the power available from the source.*

The corresponding expression is

$$G_A = \frac{P_{AVN}}{P_{AVS}} = \frac{(1-|\Gamma_S|)^2}{|1-S_{11}\Gamma_S|^2}|S_{21}|^2\frac{1}{(1-|\Gamma_{OUT}|)^2} \tag{1.86}$$

We note that $G_A$ is a function of $\Gamma_S$ while it is independent of $\Gamma_L$. So, in amplifier design (according to Fig. 1.31), the available power gain is typically of interest for low-noise amplifiers, because in such designs $\Gamma_S$ is designed to be $\Gamma_{OPT}$ in order to achieve low-noise performance (see Section 1.3.2). Usually this value is not matched to the input impedance of the central active block, and therefore the available power gain is compromised.

The next FOMs gain in importance, due to the global drive for less energy consumption. They are a measure of the efficiency, namely how efficiently the amplifier converts the supplied DC power and RF input power into RF output power. We make the distinction between efficiency ($\eta$) and power added efficiency (PAE). Note that we are referring now to the actual use of the amplifier, or, in other words, to the scheme in Fig. 1.30.

DEFINITION 1.16 *The efficiency ($\eta$) of an amplifier is defined as the ratio between the output power at the fundamental frequency and the supplied DC power.*

The corresponding expression is

$$\eta = \frac{P_L}{P_{DC}} \tag{1.87}$$

This efficiency is also called the drain efficiency if the active part consists of FETs (or derived devices such as HEMTs), or the collector efficiency if the amplifier is based on BJTs (or derived devices such as HBTs). It is named in this way because this definition does not take into account the RF input power injected in to the amplifier.

The second FOM in terms of efficiency is the power added efficiency or PAE. It is more complete than $\eta$ because it does take into account the RF input power. The definition is as follows.

DEFINITION 1.17 *The power added efficiency (PAE) of an amplifier is defined as the net increase in RF power divided by the DC power supplied.*

The corresponding expression is

$$\begin{aligned} \text{PAE} &= \frac{P_L - P_{IN}}{P_{DC}} \\ &= \frac{P_L}{P_{DC}}\left(1 - \frac{1}{G}\right) \\ &= \eta\left(1 - \frac{1}{G}\right) \end{aligned} \tag{1.88}$$

As we can deduce from Eq. (1.88), the PAE converges to $\eta$ in the case of high power gain levels.

### 1.8.2 Nonlinear FOMs

Here again, since amplifiers are two-port networks, the nonlinear FOMs as described in Section 1.5 are generally applicable. The most common FOMs that are listed in amplifier datasheets are the 1-dB-compression point $P_{1dB}$, the third-order intercept point $IP_3$, the saturated output power $P_{sat}$, the efficiency, and the power added efficiency, of which $P_{1dB}$ and $IP_3$ have already been described extensively in Section 1.5.

When considering Fig. 1.11, we see that the output power saturates at high input power. The reason, as already explained in Section 1.4, is that the output power cannot be higher than the supplied power, this being the sum of $P_{DC}$ and $P_{IN}$. Since the power-transfer characteristic at high input powers is not perfectly flat in practical cases, the saturated power is usually determined at a certain compression point, beyond the 1-dB-compression point. A typical approach is to take the output power at the 3-dB-compression point as $P_{sat}$.

### 1.8.3 Transient FOMs

The final set of FOMs is related to the transient behavior of amplifiers. It primarily applies to VGAs, which are amplifiers whose gain can be varied by means of a DC control signal. Depending on the design architecture, this control signal can be part of the DC bias supply feeding the transistors within the VGA, or can be applied to a DC bias-dependent component in the matching network (e.g., a diode). In certain cases a VGA can also be composed of a GA followed by a variable attenuator. So the gain of the GA is constant and the attenuation is being changed. Figure 1.32 illustrates how the gain changes on varying the control voltage.

In addition to the amplifier FOMs already discussed, the range of variability of the amplifier gain and the speed of change are key points when describing VGAs. The latter means that the FOMs become dependent on the dynamic behavior of the control signal. For this reason, some typical FOMs from control theory can be applied, such as the slew rate, rise time, settling time, ringing, and overshoot.

#### 1.8.3.1 Slew rate

DEFINITION 1.18 *The slew rate of an amplifier is the maximum rate of change of the output, usually quoted in volts per second.*

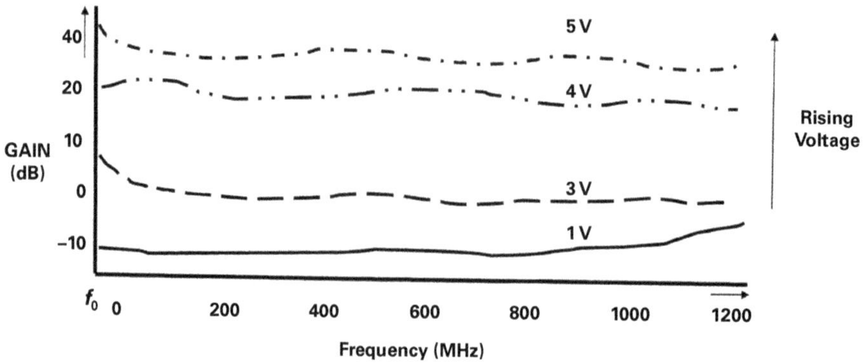


**Figure 1.32** A variable-gain amplifier controlled by a DC control bias voltage.

Many amplifiers are ultimately slew-rate-limited (typically by the impedance of a drive current having to overcome capacitive effects at some point in the circuit), which sometimes limits the full-power bandwidth to frequencies well below the amplifier's small-signal frequency response.

#### 1.8.3.2 Rise time

DEFINITION 1.19 *The rise time ($t_r$) of an amplifier is the time taken for the output to change from 10% to 90% of its final level when driven by a step input.*

#### 1.8.3.3 Settling time and ringing

DEFINITION 1.20 *The settling time is the time taken for the output to settle to within a certain percentage of the final value (for instance 0.1%).*

The next definition is ringing. Ringing is the result of overshoot caused by an under-damped circuit.

DEFINITION 1.21 *Ringing refers to an output variation that cycles above and below an amplifier's final value, and leads to a delay in reaching a stable output.*

#### 1.8.3.4 Overshoot

DEFINITION 1.22 *The overshoot is the amount by which the output exceeds its final, steady-state value, in response to a step input.*

Table 1.4 presents a typical datasheet of a VGA. The main difference from a typical amplifier datasheet is related to the range of gain variability and the speed of this change. In case of the example shown in Fig. 1.32, we see that the response time (10% to 90%) is 25 μs and the control range is 30 dB.

## 1.9 Mixers

As we can deduce from Fig. 1.1, mixers are essential components in wireless transceivers since they up-convert the modulated signal from baseband to RF for wireless transmission, and then also down-convert the received RF signal back to baseband. There are several up-conversion/down-conversion configurations (e.g., homodyne, superheterodyne, ...), among which the most important ones will be described in Chapter 2 in connection to the internal architecture of measurement instrumentation. There are also various design topologies (e.g., single-ended, balanced, double-balanced, ...), the description of which is beyond the scope of this book. The note to make in connection with FOMs, though, is that mixers can be based either on diodes (passive mixers) or on transistors (active mixers). The choice depends on the requirements of the particular transceiver design.

Figure 1.33 represents a down-converting mixer. A mixer is a three-port circuit, with as ports the radio-frequency (RF) port, the local oscillator (LO) port, and the intermediate-frequency (IF) port. The RF signal enters the RF port and gets down-converted to baseband, and then the corresponding baseband (or IF) signal exists at the IF port. The mixer is driven into nonlinear operation by a pump signal, this being

**Table 1.4** The variable-gain amplifier. The datasheet for a broadband amplifier. The electrical characteristics are response time (10% to 50%) 25 μs, and control voltage 0 to 5 V.

| Parameter | | Value | Units |
|---|---|---|---|
| Frequency | $f_L$ | 10 | MHz |
| | $f_U$ | 1200 | |
| Gain | Minimum | 24 | dB |
| | Typical | 34 | |
| | Flatness | ±1.5 | |
| | Control range | 30 | |
| Maximum power | Output 1-dB CP | +13 | dBm |
| | Input (no damage) | +10 | |
| Dynamic range | NF (typical) | 15 | dB |
| | $IP_3$ (typical) | +25 | dBm |
| VSWR | In | 2.2 | dB |
| | Out | 2.0 | |
| DC power | Voltage | 15 | V |
| | Current | 170 | mA |

Maximum ratings

| | |
|---|---|
| Operating temperature | −20 °C to 71 °C |
| Storage temperature | −55 °C to 100 °C |
| DC voltage | 17 V |

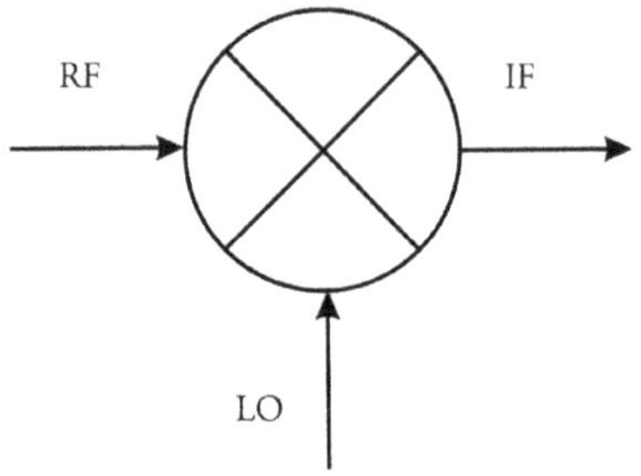


**Figure 1.33** The general mixer symbol, where the three ports are identified.

the local oscillator. In the case of an up-converting mixer, the IF port is the input and the RF port acts as the output.

Table 1.5 shows an example datasheet of a mixer. The mixer FOMs will be clarified in the next sections. We make the distinction between two-port and three-port FOMs. Throughout the FOM definitions and expressions, we assume that the mixer is a down-converter. Similar expressions for an up-converting mixer can easily be deduced.

## 1.9.1 Two-port FOMs

In terms of FOMs, a mixer is often considered as a two-port circuit, because the operation of interest is happening between the RF and IF ports. The modulated signal is

**Table 1.5** A mixer datasheet

| Parameter | | Value | Units |
|---|---|---|---|
| Frequency | RF/LO<br>IF | 250–3250<br>DC–800 | MHz |
| Conversion loss | Typical<br>Maximum | 6.5<br>8.5 | dB |
| Isolation LO/RF | Typical<br>Minimum | 30<br>15 | dB |
| Isolation LO/IF | Typical<br>Minimum | 10<br>5 | dB |
| Isolation RF/IF | Typical<br>Minimum | 30<br>15 | dB |
| LO power | Nominal | +17 | dBm |
| 1-dB-compression point | Typical | +10 | dBm |
| Input $IP_3$ | Typical | +18 | dBm |
| RF input power | Maximum | 100 | mW |
| Impedance | Nominal | 50 | Ω |
| Operating temperature | | −40 to 85 | °C |

down-converted between the RF port and IF port, or up-converted between the IF and RF ports, while the signal entering the LO port is kept constant at a high power level. For this reason, the concepts introduced in Section 1.5 are applicable to mixers.

The first FOM is the conversion loss $L$.

DEFINITION 1.23 *The conversion loss (L) of a down-converting mixer is the ratio of the RF input power and the IF output power.*

Mathematically,

$$L = \frac{P_{\mathrm{AVS,RF}}}{P_{\mathrm{AVN,IF}}} \tag{1.89}$$

where $P_{\mathrm{AVS,RF}}$ is the power of the modulated input signal at the RF carrier frequency, and $P_{\mathrm{AVN,IF}}$ is the output power of the down-converted signal at IF.

In the case of a diode-based mixer, there is always a conversion loss since diodes have no gain. Even using transistors as the nonlinear element in the mixer, there is a conversion loss or at most a small conversion gain. The reason is that the transistors have to be operated in a strongly nonlinear condition, e.g., near to pinch-off operating conditions, which corresponds to low gain. Owing to losses in the matching networks, the overall conversion is often a loss.

The FOMs which are a measure for the linearity of the mixer are the input 1-dB-compression point $\mathrm{InP}_{1\mathrm{dB}}$ and the input third-order intercept point IIP3. These two FOMs have already been introduced in Sections 1.5.1 and 1.5.2, but in the case of mixers they are usually referred to the input, as opposed to what is done in the case of amplifiers, where $P_{1\mathrm{dB}}$ and IP3 are usually referred to the output.

Mixers also have a noise contribution. The corresponding FOM is the single-sideband noise figure (SSBNF). It is defined in an analogous way to the noise figure NF of a linear two-port device (see Section 1.3.2), but now taking into account that the frequency at the output of a mixer is different from the frequency at the input of the mixer.

DEFINITION 1.24 *The single-sideband noise figure (SSBNF) for a mixer is the decibel value of the signal-to-noise ratio at the input of the mixer at RF divided by the signal-to-noise ratio at the output of the mixer at IF.*

Mathematically, the noise factor for a mixer $F_{\mathrm{mixer}}$ is

$$F_{\mathrm{mixer}} = \frac{S_{\mathrm{I,RF}}/N_{\mathrm{I,RF}}}{S_{\mathrm{O,IF}}/N_{\mathrm{O,IF}}} = \frac{N_{\mathrm{O,IF}}}{N_{\mathrm{I,RF}}L} \tag{1.90}$$

where $S_{\mathrm{I,RF}} = P_{\mathrm{AVS,RF}}$ is the signal power at the input of the mixer at RF, $S_{\mathrm{O,IF}} = P_{\mathrm{AVS,IF}}$ is the signal power at the output of the mixer at IF, $N_{\mathrm{I,RF}}$ is the available noise power at the input of the mixer at RF, $N_{\mathrm{O,IF}}$ is the available noise power at the output of the mixer at IF, and $L$ is the conversion loss.

Consequently, the noise figure is given by

$$\mathrm{SSBNF} = 10\log_{10} F_{\mathrm{mixer}} \tag{1.91}$$

### 1.9.2 Three-port FOMs

The next set of FOMs consists of FOMs that are related to the three-port nature of the mixer.

In general, a mixer has mismatches at each of its ports, like any other microwave circuit. As in the case of linear two-port networks (see Section 1.2), the port mismatches of a mixer are represented by the VSWR or return loss RL. Even though the mixer is a nonlinear circuit, and thus the reflection coefficients at each of the three ports may be dependent on the input power, a constant value corresponding to normal operating conditions is listed in mixer datasheets.

The aim of a down-converting mixer is to convert the signal at RF to a signal at IF. Any other frequency components in the signal's spectrum are unwanted, and therefore should be avoided. A distinction is made between on the one hand the leakage of the LO and IF signals to the other ports (see later) and on the other hand the harmonics and intermodulation products generated due to the nonlinear operating mode of the mixer. Owing to the small frequency offset between the RF and LO signals, the out-of-band unwanted spectral components can easily be filtered out. Commercial mixers usually have such filters included in the package already.

Since the LO power is high in order to drive the mixer in nonlinear operation, one should avoid having part of this signal leak to the RF and IF ports, because such signal may damage the circuit block preceding or following the mixer. Similarly, one wants to avoid having part of the IF output power couple back to the RF port. The possibility of leakage from the IF port to the LO port is usually ignored, since the IF power is very small compared with the LO power. In the case of passive circuits and amplifiers, such unwanted coupling between ports is called isolation (see Section 1.8.1). In mixer datasheets, both "isolation" and "leakage" are used. The various leakages are defined as follows.

DEFINITION 1.25 *The LO/RF leakage of a mixer is equal to the ratio of the power at LO frequency at the RF port and the LO power.*

Mathematically,

$$\text{LO/RF leakage} = \frac{P_{\text{RF,LOfreq}}}{P_{\text{LO}}} \tag{1.92}$$

with $P_{\text{LO}}$ the input power at the LO port and at LO frequency, and $P_{\text{RF,LOfreq}}$ the output power at the RF port and at LO frequency.

DEFINITION 1.26 *The LO/IF leakage of a mixer is equal to the ratio of the power at LO frequency at the IF port and the LO power.*

Mathematically,

$$\text{LO/IF leakage} = \frac{P_{\text{IF,LOfreq}}}{P_{\text{LO}}} \tag{1.93}$$

with $P_{\text{IF,LOfreq}}$ the output power at the IF port and at LO frequency.

DEFINITION 1.27 *The RF/IF leakage of a mixer is equal to the ratio of the power at IF frequency at the RF port and the IF power.*

Mathematically,

$$\text{IF/RF leakage} = \frac{P_{\text{RF,IFfreq}}}{P_{\text{IF}}} \tag{1.94}$$

with $P_{\text{IF}}$ the input power at the IF port and at IF, and $P_{\text{RF,IFfreq}}$ the output power at the RF port and at IF.

The final thing to note is that all of these FOMs are temperature-dependent.

## 1.10 Oscillators

### 1.10.1 Oscillator FOMs

Oscillators can come in various forms, so we will include in this chapter various of these flavors. They can be free-running oscillators, voltage-controlled oscillators, and synthesized ones. For all of those types, the oscillator characteristic to be considered as the first and most important FOM is the frequency of operation.

Actually, oscillators are produced to generate a typical signal waveform. In the case of microwave and RF circuits, this waveform is most of the time a sinusoidal signal.

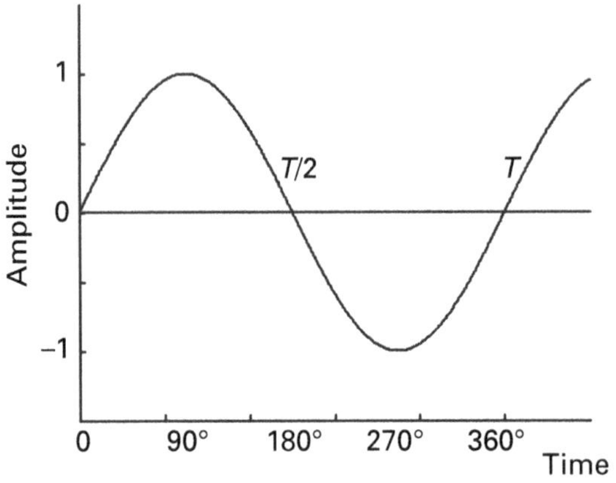


**Figure 1.34** A sine-wave oscillator, where the period definition is marked.

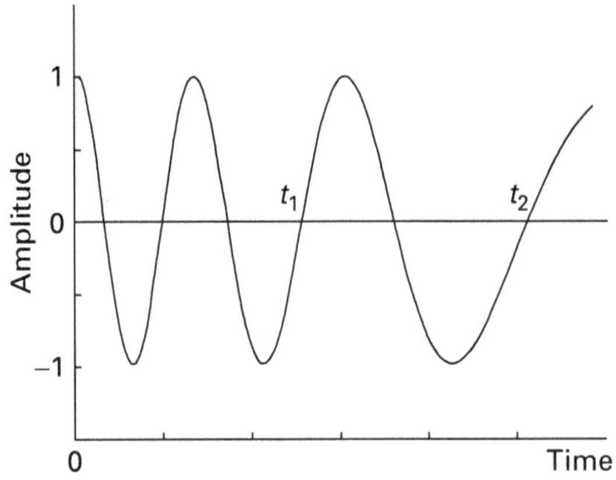


**Figure 1.35** Frequency stability, where the change in signal frequency over time can be seen.

A sine-wave signal generator produces nothing other than a voltage that changes as a function of time in a sinusoidal manner, as shown in Fig. 1.34.

In this sine wave we can define a frequency, which is the number of cycles per second at which the waveform repeats itself, and the amplitude of the waveform. Mathematically, it is represented by

$$V(t) = A \sin\left(2\pi \frac{t}{T} + \theta\right) = A \sin(2\pi f + \theta) \tag{1.95}$$

It is expected that an oscillator is a pure sine wave, but most of the time this is not true. The generator is corrupted by other factors that will have a significant impact on the output waveform. For instance, on considering Fig. 1.35, it is clear that an important characteristic of the oscillator is the frequency stability, that is how well the frequency is maintained over time. In fact the term frequency stability encompasses the concepts of random noise, intended and incidental modulation, and any other fluctuations of the output frequency of a device.

#### 1.10.1.1 Frequency stability

DEFINITION 1.28 *Frequency stability is in general the degree to which an oscillating source produces the same frequency value throughout a specified period of time. It is implicit in this general definition of frequency stability that the stability of a given frequency decreases if the wave shape of the signal is anything other than a perfect sine function.*

We can also further present the short-term and long-term stability, which are usually expressed in terms of parts per million per hour, day, week, month, or year. Long-term stability represents phenomena caused by the aging of circuit elements and of the material used in the frequency-determining element. Short-term stability relates to frequency changes of duration less than a few seconds about the nominal frequency. The reader is directed to [10] for more information.

Since we are dealing with an electronic generator, that is, one based on a strong nonlinearity, the generation of harmonics, as explained in Section 1.4.1, is also very important, since it can create harmonic components in the output signal.

#### 1.10.1.2 Phase noise

The presence of noise in the electronic components can also create another non-ideal behavior of the sine wave. This is most of the time accounted for using the FOM called phase noise (Fig. 1.36), since the noise will behave as a corrupted phase in the output signal. In this case the output sine wave can be represented as

$$V(t) = [A + \epsilon(t)]\sin(2\pi f + \phi(t)) \tag{1.96}$$

This phase-noise FOM is the term most widely used to describe the frequency stability's characteristic randomness. There are also other terms, such as spectral purity, which refers to the ratio of signal power to phase-noise sideband power.

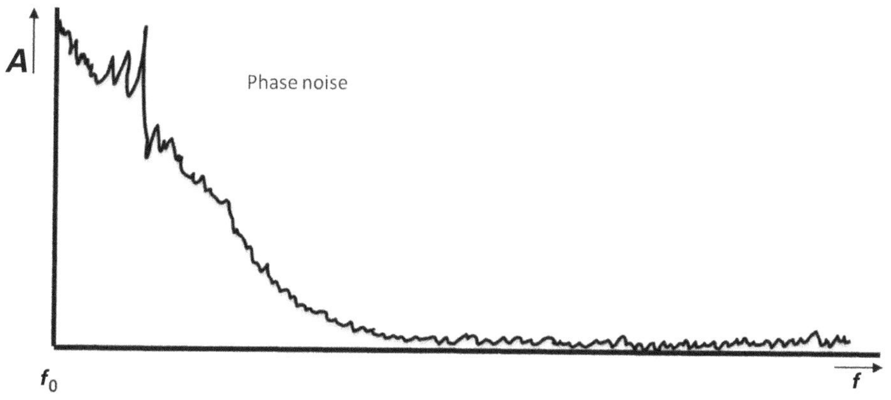


**Figure 1.36** Phase-noise representation.

**Table 1.6** An oscillator data sheet

| Parameter | Test condition | Minimum | Typical | Maximum | Units |
|---|---|---|---|---|---|
| Nominal frequency[a] | LVDS/CML/LVPECL | 10 | | 945 | MHz |
| | CMOS | 10 | | 160 | |
| Temperature stability | $T_A = -40$ to $+85\,^\circ$C | −20<br>−50<br>−100 | | +20<br>+50<br>+100 | ppm |
| Absolute pull range | ±25 | | ±345 | | ppm |
| Aging | Frequency drift over first year | | | ±3 | ppm |
| | Frequency drift over 15-year life | | | ±10 | |
| Power-up time[b] | | | | 10 | ms |

[a]Nominal output frequency set by $V_{CNOM} = V_{DD}/2$.
[b]Time from power-up or tri-state mode to $f_0$.

For a correct evaluation of frequency stability, and thus of phase noise, we should calculate the power spectral density of the waveform. In this case, for the waveform presented by Eq. (1.96), the power spectral density is

$$G_{\text{sideband}} = \int_{-\infty}^{+\infty} S_g(f)df \tag{1.97}$$

where $S_g(f)$ represents the two-sided spectral density of fluctuations of the output waveform.

Actually Glaze, in his chapters in [10], discussed a possible definition of frequency stability that relates the sideband power of phase fluctuations to the carrier power level. This quantity is called $\pounds(f)$.

DEFINITION 1.29 *$\pounds(f)$ is defined as the ratio of the power in one sideband, referred to the input carrier frequency on a per-hertz-of-bandwidth spectral-density basis, to the total signal power, at Fourier frequency difference $\delta f$ from the carrier, per device. In fact, it is a normalized frequency-domain measure of phase-fluctuation sidebands, expressed in decibels relative to the carrier per hertz:*

$$\pounds(f) = \frac{\text{power density (one phase-modulation sideband)}}{\text{carrier power}} \tag{1.98}$$

On looking at a typical oscillator datasheet from a manufacturer, it is clear that most of the FOMs explained here are represented in the datasheet, as depicted in Table 1.6.

## 1.11 Frequency-multiplier FOMs

Frequency multipliers are circuits that convert an input signal at frequency $f_0$ into an output signal at a frequency that is a multiple of $f_0$. Their use in wireless transceivers is usually in combination with oscillators. The higher the required LO frequency, the more

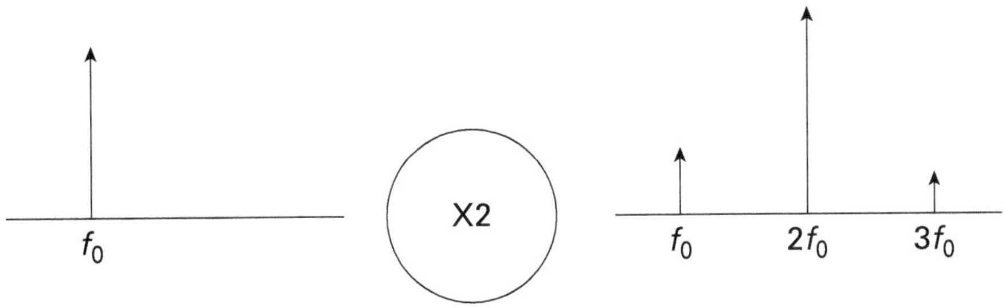


**Figure 1.37** Frequency doubling, showing the input and output spectrum components.

difficult it is to design and fabricate oscillators with low phase noise. So the approach used is to take a very good oscillator at a lower LO frequency and then up-convert this frequency by means of a frequency multiplier. Typical multiplication factors in practical designs are in the range 2–4, because the higher the multiplication factor the higher the conversion loss (see later for the definition). Using a frequency multiplier does increase the phase noise, by a factor equal to the multiplication factor, but the resulting phase noise is still typically lower than that of an oscillator at the higher frequency.

Figure 1.37 shows schematically a frequency doubler. In this example, the input signal is a single-tone signal at frequency $f_0$, and the intended output signal is at $2f_0$. Since the output signal of a linear circuit is by definition at the same frequency as the frequency of the input signal (see also Section 1.2), any frequency-converting circuit (frequency multipliers but also mixers) is a nonlinear circuit, and therefore unwanted spectral components are generated at the output as well, e.g., at $f_0$ and $3f_0$. The latter are characterized by the frequency-multiplier FOMs which we will define next.

The first FOM of a frequency multiplier is the conversion loss, which is defined similarly to that in the case of mixers (see Section 1.9.1). The definition, with $n$ the multiplication factor, is as follows.

DEFINITION 1.30 *The conversion loss (L) of an n times frequency multiplier is the ratio of the input power at frequency $f_0$ and the output power at frequency $nf_0$.*

Mathematically,

$$L = \frac{P_{\mathrm{AVS},f_0}}{P_{\mathrm{AVN},nf_0}} \tag{1.99}$$

As in the case of mixers, there is usually no conversion gain, even when the frequency multiplier is based on transistors.

At the output, we want to have a clear output signal at the multiplied frequency. So both the $f_0$ spectral component and unwanted harmonics ($mf_0|_{m\neq n}$) should be suppressed. This is achieved by design, e.g., by using a circuit architecture that automatically suppresses the $f_0$ component, and/or by incorporating dedicated filters in the circuit's package. The corresponding FOMs are the fundamental and harmonic rejections.

**Table 1.7** A frequency-multiplier datasheet

| Multiplication factor | Frequency (GHz) $f_1$ In | $f_2$ Out | Input power Min. | Max. | Conversion loss Typ. | Max. | Harmonic output $f_1$ Typ. | $f_1$ Min. | $f_3$ Typ. | $f_3$ Min. | $f_4$ Typ. | $f_4$ Min. |
|---|---|---|---|---|---|---|---|---|---|---|---|---|
| 2 | 5–8 | 10–16 | 13 | 16 | 11.5 | 15 | 30 | 18 | 35 | 23 | 25 | 15 |
| | 8–10 | 16–20 | 10 | 13 | 15 | 18 | 21 | 14 | 30 | 18 | 20 | 13 |
| | | | 13 | 16 | 12 | 15 | 33 | 20 | 27 | 17 | 50 | 35 |
| | | | 10 | 13 | 15 | 18.5 | 30 | 16 | 23 | 16 | 40 | 30 |

Min., minimum; Max., maximum; Typ., typical.

DEFINITION 1.31 *The fundamental rejection is the ratio of the output power at frequency $nf_0$ and the output power at frequency $f_0$.*

Mathematically,

$$\text{fundamental rejection} = \frac{P_{\text{AVN},nf_0}}{P_{\text{AVN},f_0}} \tag{1.100}$$

The rejection is usually expressed in dBc, which is the difference, expressed in decibels, relative to the wanted signal, or carrier.

DEFINITION 1.32 *The harmonic rejection is the ratio of the output power at frequency $nf_0$ and the output power at frequency $f_{m0}|_{m\neq n\neq 0}$.*

Mathematically,

$$\text{harmonic rejection} = \frac{P_{\text{AVN},nf_0}}{P_{\text{AVN},mf_0}\big|_{m\neq n\neq 0}} \tag{1.101}$$

Finally, as with all electronic circuits, the characteristics of a frequency multiplier are temperature dependent.

An example datasheet of a frequency doubler is presented in Table 1.7.

## 1.12 Digital converters

Finally, this chapter would not be complete if digital converters were not included.

Digital converters, either analog to digital (ADC) or digital to analog (DAC), are becoming a key component in radio and wireless communication circuits and systems. Actually, the advent of software-defined radio (SDR) and/or cognitive radio (CR) is moving this technology faster to higher frequencies and thus to new characterization procedures.

An ADC converts an analog signal to digital quantities, by operation in two axes over the continuous time-domain analog signal. These two axes correspond to a sampling in time and a sampling in amplitude, usually called quantization [11].

Sampling in time corresponds to sampling the time-domain signal at some discrete points with a sampling frequency that should obey the Nyquist frequency. Sampling in

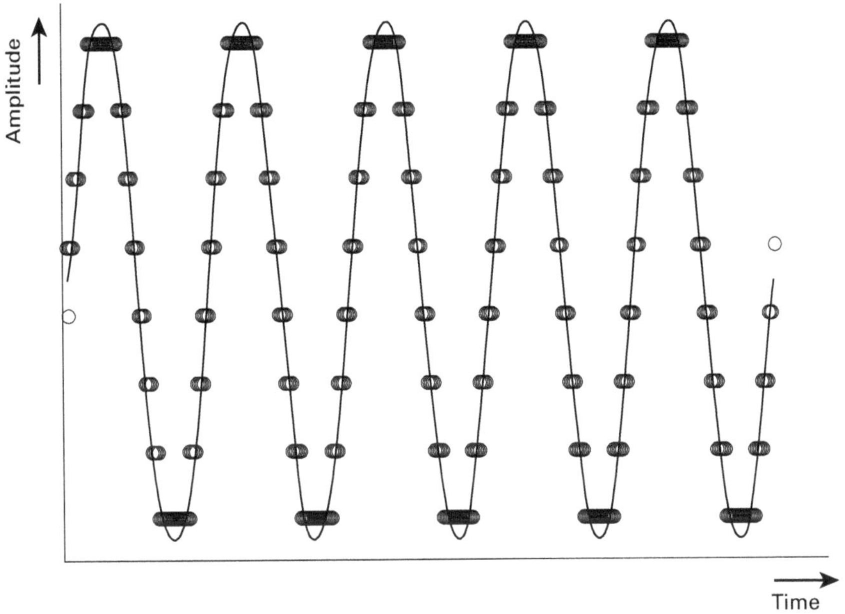


**Figure 1.38** Sampling and quantization procedures.

amplitude corresponds to sampling the amplitude axis at discrete points also. Quantization and sampling at the same time corresponds to picking up the sampled points both in time and amplitude and constraining them to fit a pre-determined matrix. This fact leads to several important FOMs, since the time sample can and will generate aliasing errors, while the quantization will generate a minimum-noise floor that will severely degrade the output digital signal, by adding quantization noise to it. This is illustrated in Fig. 1.38.

The study of sampling and quantization noise is beyond the scope of this book, but the reader is directed to reference [11] for more information.

Traditional ADC/DAC applications are at low frequencies, and thus some of the characterizations and FOMs used are low-frequency-related ones. For instance, we can have offset errors, gain errors, integral nonlinearities, differential nonlinearities, and special FOMs related to the quantization noise.

### 1.12.1 Figures of merit

The more traditional FOMs can be defined as follows.

DEFINITION 1.33 *The gain error is the difference between the measured and ideal full-scale input voltage range of the ADC.*

DEFINITION 1.34 *The offset error is the DC offset imposed on the input signal by the ADC, reported in terms of LSB (codes).*

### 1.12.1.1 ADC time behavior

Some of the FOMs related to the time behavior of the ADC include the following:

DEFINITION 1.35 *The aperture uncertainty is related to the signal jitter, which is the sample-to-sample variation in aperture delay.*

DEFINITION 1.36 *The Encode pulse width/duty cycle. Pulse width high stands for the minimum amount of time the Encode pulse should be left in the Logic 1 state to achieve the rated performance, while pulse width low is the minimum time the Encode pulse should be left in the low state.*

DEFINITION 1.37 *The maximum conversion rate is the maximum Encode rate at which the image spur calibration degrades by no more than 1 dB.*

DEFINITION 1.38 *The minimum conversion rate is the minimum Encode rate at which the image spur calibration degrades by no more than 1 dB.*

DEFINITION 1.39 *The output propagation delay is the delay between a differential crossing of one Encode and another Encode (or zero crossing of a single-ended Encode).*

DEFINITION 1.40 *The pipeline latency is the number of clock cycles by which the output data lags relative to the corresponding clock cycle.*

DEFINITION 1.41 *The signal-to-noise ratio for ADCs ($SNR_{ADC}$) is the ratio of the RMS signal amplitude (set at 1 dB below full scale) to the RMS value of the sum of all other spectral components, excluding harmonics and DC.*

For an ideal ADC one can represent this value as

$$\mathrm{SNR_{ADC}} = 6.02N + 1.76\,\mathrm{dB} \tag{1.102}$$

where $N$ is the number of bits of the ADC.

DEFINITION 1.42 *The effective number of bits (ENOB) corresponds to the number of bits that we can have when considering not the ideal but the measured SNR.*

The ENOB is calculated from the measured SNR as follows:

$$\mathrm{ENOB} = \frac{\mathrm{SNR_{MEASURED}} - 1.76}{6.02} \tag{1.103}$$

### 1.12.1.2 ADC nonlinear behavior

Some FOMs are also related to the nonlinear behavior of the ADC.

DEFINITION 1.43 *The differential nonlinearity is the type of nonlinearity that corresponds to the deviation of any code width from an ideal one-least-significant-bit (LSB) step.*

DEFINITION 1.44 *The integral nonlinearity is the deviation of the transfer function from a reference line measured in fractions of 1 LSB using a best straight line determined by a least-square-curve fit.*

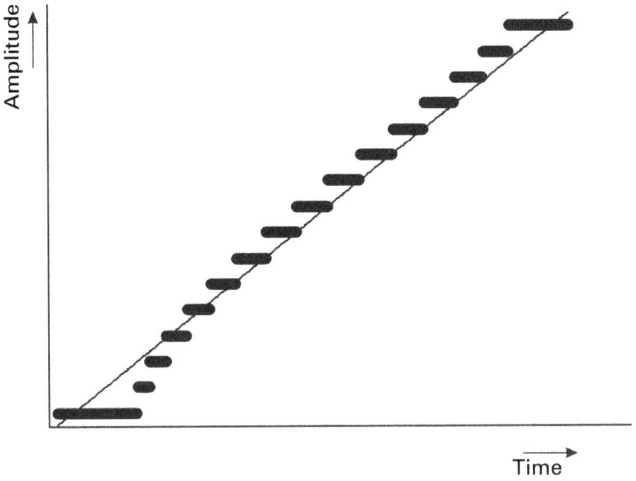


**Figure 1.39** The impact of nonlinear behavior on the ADC quantization curves.

These two FOMs can be seen in Fig. 1.39.

From a wireless point of view, nonlinearity can also be characterized by using typical measures of intermodulation and harmonics as previously presented in Section 1.5.

DEFINITION 1.45 *The second-harmonic distortion is the ratio of the RMS signal amplitude to the RMS value of the second-harmonic component, reported in dBFS (dBFS means dB full scale, which means x dB below the full scale).*

DEFINITION 1.46 *The third-harmonic distortion is the ratio of the RMS signal amplitude to the RMS value of the third-harmonic component, reported in dBFS.*

DEFINITION 1.47 *The signal-to-noise-and-distortion (SINAD) ratio is the ratio of the RMS signal amplitude (set 1 dB below full scale) to the RMS value of the sum of all other spectral components, including harmonics, but excluding DC and image spur.*

DEFINITION 1.48 *The two-tone intermodulation distortion rejection is the ratio of the RMS value of either input tone to the RMS value of the worst third-order intermodulation product, reported in dBc.*

Sometimes the characterization of analog converters also includes several FOMs related to any spurious signals that are visible at the output signal. Those FOMs include the following.

DEFINITION 1.49 *The spurious-free dynamic range (SFDR) is the ratio of the RMS signal amplitude to the RMS value of the peak spurious spectral component, except the image spur. The peak spurious component may, but need not, be a harmonic. It can be*

**Table 1.8** An ADC datasheet (for $V_A = 3.7$ V, $V_C = 1.5$ V, ENCODE = 400 MSPS, and $0\,°C \le T \le 60\,°C$, unless specified otherwise)

| Parameter | | Case $T$ | Minimum | Typical | Maximum | Units |
|---|---|---|---|---|---|---|
| Dynamic performance | | | | | | |
| SNR | | | | | | |
| Analog input | 10 MHz | Full | 62 | 64 | | dBFS |
| | 70 MHz | Full | 61.5 | 63.5 | | dBFS |
| at −1.0 dBFS | 128 MHz | Full | 61.5 | 63.5 | | dBFS |
| | 175 MHz | Full | 61.5 | 63.5 | | dBFS |
| SINAD ratio | | | | | | |
| Analog input | 10 MHz | Full | 59 | 63.5 | | dBFS |
| | 70 MHz | Full | 58.5 | 63 | | dBFS |
| at −1.0 dBFS | 128 MHz | Full | 57.5 | 61.5 | | dBFS |
| | 175 MHz | Full | 55 | 60 | | dBFS |
| SFDR | | | | | | |
| Analog input | 10 MHz | Full | 69 | 85 | | dBFS |
| | 70 MHz | Full | 69 | 80 | | dBFS |
| at −1.0 dBFS | 128 MHz | Full | 66 | 72 | | dBFS |
| | 175 MHz | Full | 62 | 68 | | dBFS |
| Image Spur | | | | | | |
| Analog input | 10 MHz | Full | 60 | 75 | | dBFS |
| | 70 MHz | Full | 60 | 72 | | dBFS |
| at −1.0 dBFS | 128 MHz | Full | 60 | 66 | | dBFS |
| | 175 MHz | Full | 57 | 63 | | dBFS |
| Offset spur | | | | | | |
| Analog input at −1.0 dBFS | | 60 °C | | 65 | | dBFS |
| Two-tone IMD | | | | | | |
| $F_1$, $F_2$ at −6 dBFS | | 60 °C | | −75 | | dBc |
| Analog input | | | | | | |
| Frequency range | | Full | 10 | | 175 | MHz |
| Digital Input $DR_EN$ | | | | | | |
| Minimum time, low | | Full | 5.0 | | | ns |
| Switching specifications | | | | | | |
| Conversion rate | | Full | 396 | 400 | 404 | MSPS |
| Encode pulse width high | | 60 °C | | 1.25 | | ns |
| Encode pulse width low | | 60 °C | | 1.25 | | ns |

*reported in dBc (that is, it degrades as the signal level is lowered) or dBFS (always related back to converter full-scale).*

Other FOMs consider the VSWR, IMR, ACPR, NPR, etc. which were presented in Sections 1.3 and 1.5 on linear and nonlinear FOMs.

Table 1.8 presents a typical ADC datasheet. From the datasheet it is clear that the FOMs of an ADC related to the RF part are focused specifically on the signal-to-noise ratio (SNR), or, if we include distortion, the signal-to-noise-and-distortion (SINAD) ratio, or, if we include all types of possible spurious we can have the spurious–free dynamic range, which is related not only to the noise, but also to the maximum signal allowed before clipping. All these characteristics are expressed in dBFS.

## Problems

**1.1** Prove the expressions for $\Gamma_{IN}$ and $\Gamma_{OUT}$ in Eq. (1.82).

**1.2** Knowing that a receiver has a sensitivity of −100 dBm, what should be the maximum power an out-of-band interferer can have in order to maintain signal quality? Consider that the receiver has a gain of 10 dB, that the useful signal has a bandwidth of 100 KHz, an NF of 2 dB, 1-dB-compression point of 30 dBm, and an $IP_3$ of 47 dBm, and that signals out of band have an extra 50 dB rejection due to the use of an input filter.

**1.3** In your view, what is preferable in a transceiver: to have high gain and low NF at the first stage of the receiver, or to have high gain and high $IP_3$?

**1.4** In a receiver chain, the first block is a cable, followed by an amplifier and a mixer. Considering that the loss in the cable is 3 dB, the amplifier gain is 10 dB with an NF of 2 dB, and the mixer has an insertion loss of 10 dB, calculate the overall NF.

**1.5** Explain why we use a low-noise amplifier in the receiver chain and a power amplifier in the transmitter chain.

**1.6** What is the main impact of phase noise?

**1.7** How can you evaluate the impact of phase noise on digital systems?

**1.8** What is the difference between gain and underlying linear gain?

**1.9** Comment on the relationship between the 1-dB-compression point and $IP_3$.

**1.10** Explain why the EVM can be calculated using the SNR, not the ACPR.

**1.11** What are the main limitations of digital converters?

**1.12** If we have a system with NPR = 10 dB, what will the EVM be?

## References


[1] M. B. Steer, *Microwave and RF Design: A Systems Approach*. Herndon, VA: SciTech Publishing, 2010.

[2] D. M. Pozar, *Microwave Engineering*. New York: John Wiley, 2005.

[3] S. Maas, *Noise in Linear and Nonlinear RF and Microwave Circuits*. Norwood, MA: Artech House, 2005.

[4] J. C. Pedro and N. B. Carvalho, *Intermodulation Distortion in Microwave and Wireless Circuits*. New York: Artech House, 2003.

[5] S. Maas, *Nonlinear Microwave and RF Circuits*. Norwood, MA: Artech House, 2003.

[6] N. B. Carvalho and J. C. Pedro, "A comprehensive explanation of distortion side band asymmetries," *IEEE Trans. Microwave Theory Tech.*, vol. 50, no. 9, pp. 2090–2101, Sep. 2002.

[7] N. Carvalho, K. Remley, D. Schreurs, and K. Gard, "Multisine signals for wireless system test and design," *IEEE Microwave Mag.*, vol. 9, no. 3, pp. 122–138, Jun. 2008.

[8] K. Gharaibeh, K. Gard, and M. Steer, "Accurate estimation of digital communication system metrics – SNR, EVM and $\varrho$ in a nonlinear amplifier environment," in *ARTFG Microwave Measurements Conference*, Honolulu, Hawaii, Jun. 2004.
[9] F.-L. Luo, *Digital Front-End in Wireless Communications and Broadcasting, Circuits and Signal Processing*. Cambridge: Cambridge University Press, 2011.
[10] D. Sullivan, D. Allan, D. Howe, and F. Walls (eds.), *Characterization of Clocks and Oscillators*. Boulder, CO: National Institute of Standards and Technology, 1990.
[11] H. H. Nguyen and E. Shwedyk, *A First Course in Digital Communications*. Cambridge: Cambridge University Press, 2009, Chapter 4.

# 2 Instrumentation for wireless systems

## 2.1 Introduction

In the first chapter the main figures of merit for microwave and wireless circuits and systems were presented. Those figures of merit are fundamental for a correct specification of the system to be built, and thus their identification and measurement are key to the success of the wireless engineer.

In order to perform correct measurements of those figures of merit, a set of instruments should be used. These instruments should be jointly capable of identifying the most important parameters of the system and clearly, and without any ambiguity, capturing the correct values to be measured.

Several types of instrumentation were developed for gathering all these values and thus clearly identifying the quantities to be measured and specified. In this chapter we present a description of how these instruments work, and their main drawbacks will be explained. For each instrument the quantities to be measured, the internal architecture, the definition of the main instrument parameters, and its calibration procedure will be covered. This is done for the main and most important instruments available to microwave and wireless engineers, namely

(1) power meters
(2) spectrum analyzers
(3) vector signal analyzers
(4) real-time signal analyzers
(5) vector network analyzers
(6) nonlinear vector network analyzers
(7) oscilloscopes
(8) logic analyzers
(9) noise-figure meters

## 2.2 Power meters

Power and energy are the most important quantities to be measured in any wireless system. It is the use of power quantities that allows us to quantify and design high-quality and efficient wireless systems and networks. Actually, the two main quantities to be measured in terms of a wireless link are the transmitted power and the sensitivity (the minimum power that the system should receive for a predetermined signal-to-noise ratio). For instance, it should be clear that, in a wireless link with line of sight, doubling the power means increasing the geographic area which will be covered, and consequently more users will be served by a wireless communication system, implying a significant reduction in cost.

Actually, the demand for power is so high that a correct measurement of its value is fundamental, not only during the design of power amplifiers, but also during installation and throughout their use, for instance, a mobile phone is continuously measuring power for handover decisions.

This brings us to the discussion of how to measure the power of a wireless signal. Power is actually a physical term that is used to describe the average amount of energy, or work, that is spent per unit of time. So power is nothing other than the amount of joules spent per unit time, which is measured in watts. In the past it was also described in terms of horsepower, by analogy to steam-driven machines in the industrial revolution [1].

From a mathematical point of view the main objective when evaluating power is to calculate the amount of energy spent per time as

$$P = \frac{\text{Energy}}{\Delta t} \tag{2.1}$$

where Energy is the energy being expended during the time span $\Delta t$.

The unit of power that is used in this book is the watt, $W = J/t$ with $J$ being energy in joules and $t$ time in seconds.

From a signal perspective, energy can be calculated by integrating each individual energy pattern over time, that is

$$E_{\text{s}} = \langle x(t), x(t) \rangle = \int_{-\infty}^{\infty} |x(t)|^2 \, dt \tag{2.2}$$

with

$$E = \frac{E_{\text{s}}}{R}$$

where $E_{\text{s}}$ is the signal energy, $\langle \cdot \rangle$ represents the averaging operation over time, and $E$ represents the true energy spent on an electrical component with load $R$.

The power within a time window can thus be calculated as

$$P_{\text{s}} = \frac{1}{\tau} \int_{-\tau/2}^{\tau/2} |x(t)|^2 \, dt \tag{2.3}$$

where $\tau$ in this case is the duration of the time window. If the signal is periodic and $\tau = T$, the average power of the device is obtained.

In an electrical circuit, the energy is actually calculated as the amount of current traversing a component multiplied by the voltage across its terminals. This can be easily understood since voltage is expressed in joules per coulomb, $V = [J/C]$. The coulomb is defined as the unit of electrical charge, which corresponds to the charge transported

**Table 2.1** Logarithmic conversions

| W | dBW | dBm |
|---|---|---|
| 1 mW | −30 dBW | 0 dBm |
| 10 mW | −20 dBW | 10 dBm |
| 100 mW | −10 dBW | 20 dBm |
| 1 W | 0 dBW | 30 dBm |
| 10 W | 10 dBW | 40 dBm |

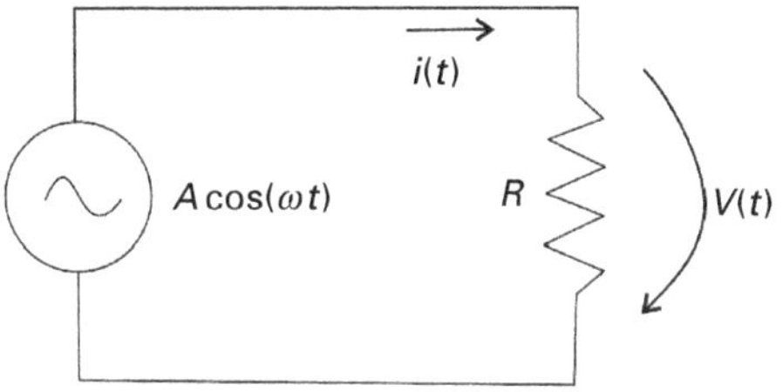


**Figure 2.1** A simple resistive circuit, used as an example for the power calculation.

by a constant current of 1A. This means that the energy and subsequently power can be given by

$$P_s = \frac{1}{\tau}\int_{-\tau/2}^{\tau/2}[v(t)i(t)]dt \tag{2.4}$$

where $v(t)$ and $i(t)$ are, respectively, the instantaneous voltage across and the instantaneous current through the electrical component.

For example, if we analyze a simple circuit consisting of a resistance $R$, as in Fig. 2.1, excited by an RF signal, for instance a sinusoidal waveform, described by its voltage over time as $v(t) = A\ \cos(\omega t)$, the average power that is dissipated in the resistance $R$ over a period of time is given by

$$P_R = \frac{1}{T}\int_{-T/2}^{T/2}[v(t)i(t)]dt = \frac{1}{T}\int_{-T/2}^{T/2}\frac{v(t)^2}{R}dt = \frac{1}{T}\int_{-T/2}^{T/2}\frac{[A\ \cos(\omega t)]^2}{R}dt = \frac{A^2}{2R} \tag{2.5}$$

Often a logarithmic scale (Table 2.1) is used in wireless communications to express power. So power is not expressed in watts, but is given in dBW, which is equal to $P\ [\text{dBW}] = 10\ \log(P[W]/1[W])$. Even more commonly, the value is referred to 1 mW, called dBm, $P\ [\text{dBm}] = 10\ \log(P[W]/1[mW])$.

## 2.2.1 How to measure power

As explained, power is the measure of energy per unit time, which means that measuring power actually corresponds to measuring energy. So power in wireless circuits can be measured indirectly using ways to measure the current and voltage at the output terminal of the wireless circuit, which is the process used for low-frequency components. At high

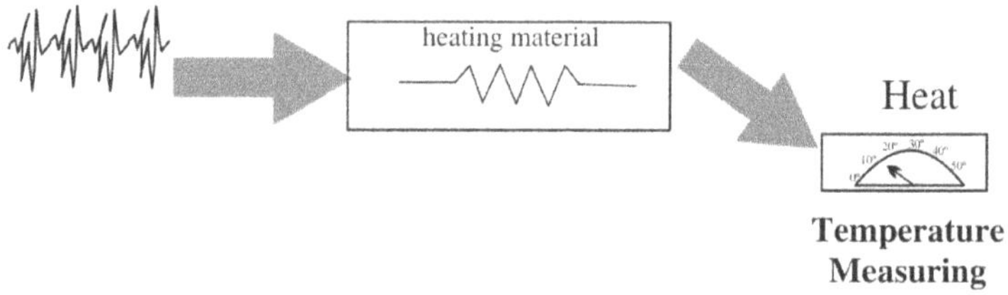


**Figure 2.2** Indirect measurement of power, using the temperature rise of a material that heats up in consequence of power dissipation.

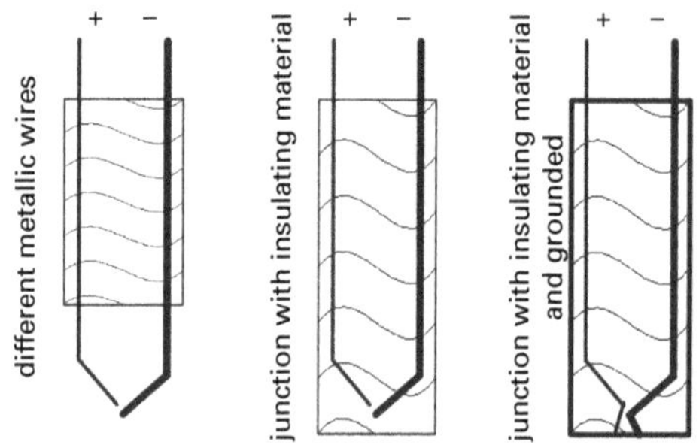


**Figure 2.3** Thermal sensors, from left to right: different metallic wires connected, using a junction with insulating material, and using a junction with insulating material and grounded.

frequencies, it is difficult to measure the voltage and current, since their values depend on the location (see Eq. (1.2)), and therefore other forms of measuring power should be considered.

Going back in time, history reveals to us that being able to measure power was actually the first important need for wireless engineers, since it was a way to guarantee that the wireless system was working correctly. Mainly the measurement of power versus frequency, giving rise to the first evaluations of the power spectral density was the interest. Nevertheless, the need in this case is still to measure a propagating wave and to evaluate either its instantaneous power, most of the time called the *peak power*, or its average value, which is normally called simply the power.

Wireless transmitted power was then measured using indirect ways; that is, the engineer tried to evaluate the power being transmitted in term of the impact on a certain material, and thereby to measure indirectly the power that most of the time was absorbed by the measuring instrument. Some of these procedures include the use of a fluorescent screen, which reveals the power absorbed by the fluorescent probe [2].

Others used the fact that sending energy into certain materials increases their temperature, and thus, by measuring the temperature rise, power can be measured indirectly (Fig. 2.2). These materials include liquids and other types of devices, Fig. 2.3 [2, 3].

Electrical probes were used as well, resulting in improved measurement accuracy. Those include thermistors and thermocouple sensors, which were very important in the 1970s (see Fig. 2.3).

More recently also diodes were used [2, 3]. Exploiting their nonlinear behavior, the DC value arising from nonlinear conversion is another indirect way to measure power. The diode probe is actually the most used approach for peak power measurements [2, 4]. Despite the fact that there are several power-meter configurations, in this book we will exclusively deal with thermocouple and diode probe power meters, which are the most important power probes on the market for wireless circuits and systems.

## 2.2.2 The thermocouple principle

As described in the previous section, thermocouples allow one to measure wireless power by evaluating indirectly the heating of a device, namely the thermocouple. Thermocouple sensors have been the optimal choice for power measurements ever since their appearance in the 1970s. The most important facts justifying this choice can be summarized as follows.

1. A thermocouple exhibits a square law that is inherent to its characteristics, meaning that the DC value at its output is proportional to the RF power, and thus to the square of the RF input voltage.
2. At that time (during the 1970s), thermocouples were the best technology, in comparison with other, similar thermal sensors.

Thermocouples are truly "average detectors," since the temperature will increase irrespective of the type of signal at their input, or, in other words, they are mainly sensitive to the average power of the exciting signal. A thermocouple can measure power levels as low as −30 dBm, and is very well behaved in terms of uncertainty due to its good values of VSWR.

But a question arises, namely what are thermocouples?

*Thermocouples are based on the fact that two different metals can generate a voltage due to temperature differences between the two metals. The physical principle is illustrated by Fig. 2.4. When a metal is heated at one terminal, several electrons become free due to thermal agitation, and the increased density of those electrons at the heating spot creates a current diffusion to the other side of the material. This current phenomenon gives rise to a force according to Coulomb's law. This force creates an electric field that will further give rise to a voltage source, called the Thomson electromotive force (emf).*

*Now, if two metals are used as in Fig. 2.5, then the same phenomenon appears, and gives rise to a diffusion of current, and subsequently to emf. This phenomenon is called the Peltier effect [2].*

*A thermocouple is thus a circuit consisting of two materials at different temperatures that, when kept together, will generate a current flowing in the loop as long as the two materials are held at different temperatures. If now the loop is broken and a voltmeter is included, then the generated emf can be measured. The thermocouple circuit will thus use both the Thomson emf and the Peltier emf to produce the final thermoelectric voltage, which is called the Seebeck emf.*

These thermocouples are cascaded to create a thermopile, which significantly increases the voltage drop. In order to heat one of the metals, the RF signal is passed

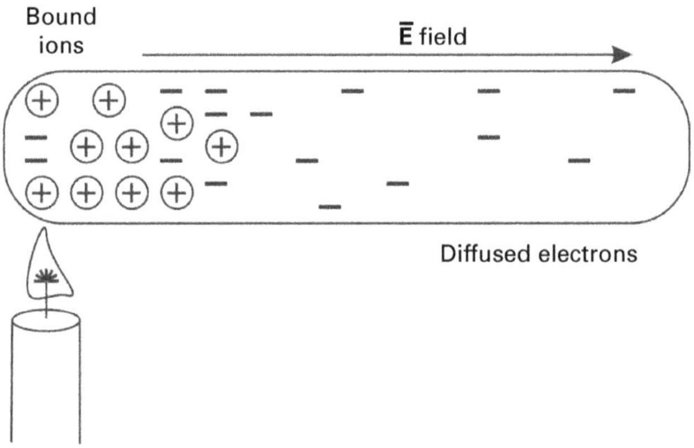


**Figure 2.4** The physical principle of a thermocouple, where a diffusion current is formed due to thermal agitation.

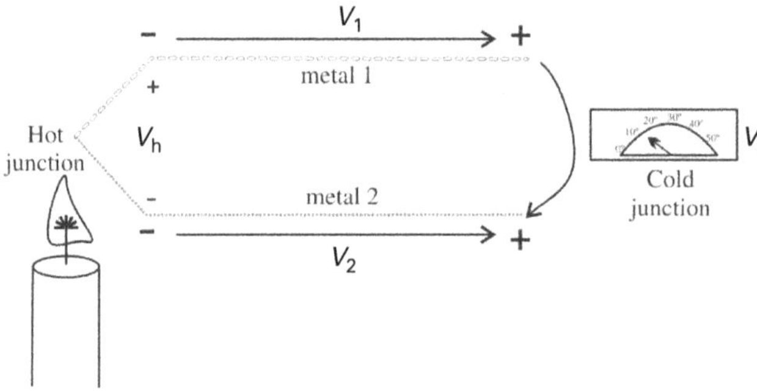


**Figure 2.5** A thermocouple in real operation, combined with a voltmeter; $V_0 = V_1 + V_h - V_2$.

through a metal resistor, normally made of thin film (usually tantalum nitride), while the sensor itself is made of silicon. A detailed analysis can be found in [2].

Therefore the complete thermocouple is actually a combination of a thermal resistor, which heats up when traversed by an RF signal, and a thermocouple that converts a temperature rise into a voltage.

A typical thermocouple sensor from a commercial manufacturer can be seen in Fig. 2.6, where two thermocouples are seen connected together to form the power sensor.

Typical sensors from manufacturers are designed in order to have a thermocouple with a predetermined thermal power, for instance 250 μV/°C, and a certain value of the thermal resistance, for instance 0.4 °C/mW. The overall sensitivity is the combination of these two values, and, in this example, it is 100 μV/mW.

Since these thermocouples are based on the heating of a thermal resistor, and the circuit is not free of thermal capacitance, the heating mechanisms will take a certain amount of time to heat the thermocouple to a steady state, and therefore it takes a certain amount of time to guarantee a precise constant voltage value (Fig. 2.7).

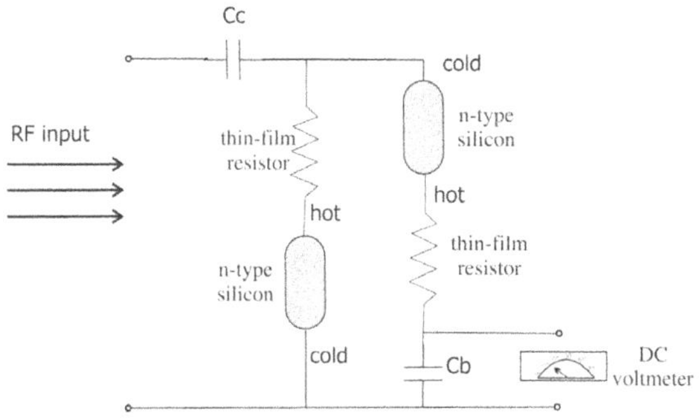


**Figure 2.6** An example of conditioning of a thermocouple signal, from a real implementation.

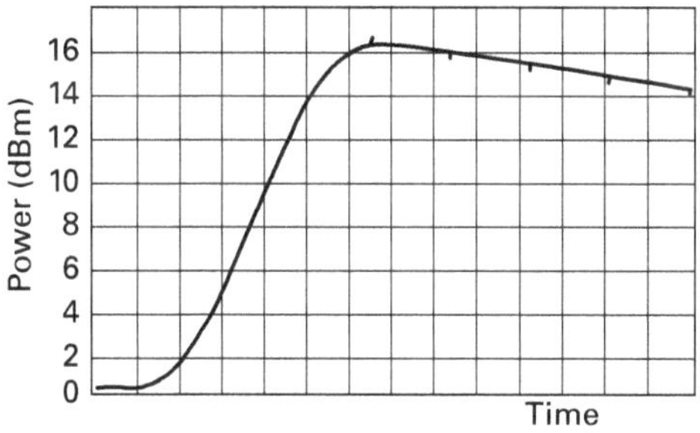


**Figure 2.7** Thermocouple measurement variation over time.

Moreover, care should be taken when using this type of sensor, mainly due to the maximum-temperature limitation and the degradation over time, especially regarding the thermal characteristics of resistors.

### 2.2.3 The diode probe principle

As we saw in Section 2.2, the objective of measuring power implies nothing other than measuring the square of the voltage over a resistor as in Eq. (2.5). Thus any element that behaves quadratically with the applied RF voltage could be a candidate for power-measurement probes.

One such element is the diode, since the diode behaves as a quadratic device in a certain part of its characteristic. Actually, one started to use diodes as power probes in the 1970s. Diodes based on the low-barrier Schottky technology were the first components at microwave frequencies that were rugged and gave repeatable results from diode to diode.

Diode probes can detect and measure power as low as −70 dBm (100 pW) at microwave frequencies.

In order to understand the basic principle of a diode power probe, consider its $I$–$V$ characteristic, as presented in Fig. 2.8.

The $I$–$V$ characteristic of the diode can be described by

$$I_{\mathrm{D}} = I_{\mathrm{s}}(e^{(V_{\mathrm{D}}/V_T)} - 1) \qquad (2.6)$$

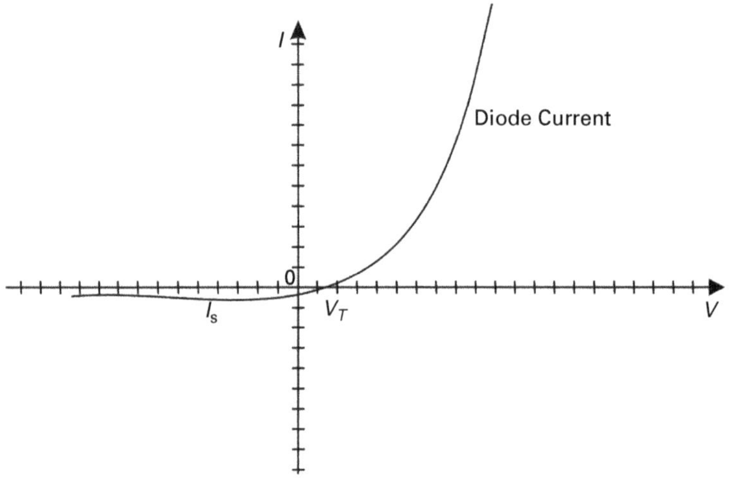


**Figure 2.8** A typical diode $I-V$ characteristic.

where $V_T = nk_B T/q$, $k_B$ is Boltzmann's constant, $T$ is the absolute temperature, $q$ is the charge of an electron, $n$ is a correction constant to fit the model to experimental data ($n$ is most of the time equal to 1.1), and $I_s$ is the inverse current.

This model presents an exponential behavior; that is, the current traversing the diode can be related to the exponential voltage applied to its terminal.

If we now approximate this function near a predetermined bias point, $V_{D_{Bias}}$, with a Taylor-series expansion, then the current flowing through the diode becomes

$$I_D = I_{D0} + K_1 v_d + K_2 v_d^2 + K_3 v_d^3 + \cdots \tag{2.7}$$

where $v_d = V_D - V_{D_{Bias}}$.

For a real circuit as sketched in Fig. 2.9(a), the variation of the diode current, and subsequently of the voltage across the resistor, $R_{LOAD}$, can be described as presented in Fig. 2.9(b). In this figure, it is possible to see that, during a certain variation of the input voltage swing, the output actually behaves quadratically. This means that the probe can be used for measurement of RF power if the input power does not exceed the range of validity. If it goes beyond that limit, then there will be an error in the measurement due to the non-quadratic behavior of the diode. However, this can be calibrated out, as we will see in Section 2.2.5.

The power-measurement mechanism in the diode power probe can be described as follows. Consider an input voltage of $V_{in} = A\cos(\omega t)$. The output of Eq. (2.7) when truncated to the second order (quadratic behavior) is given by

$$\begin{aligned} I_D &= I_{D0} + K_1 V_d + K_2 V_d^2 \\ &= I_{D0} + K_1[A\cos(\omega t)] + K_2[A\cos(\omega t)]^2 \\ &= I_{D0} + K_1[A\cos(\omega t)] + \frac{K_2}{2}[A^2\cos(2\omega t) + A^2] \end{aligned} \tag{2.8}$$

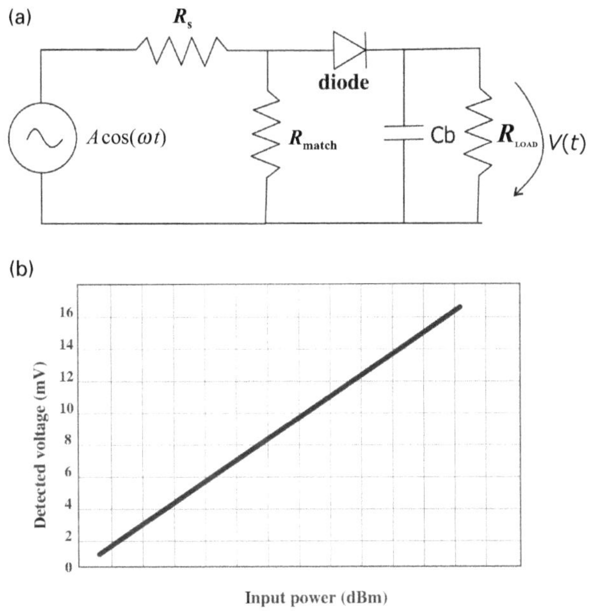


**Figure 2.9** A simple diode power probe circuit and diode quadratic behavior. (a) A simple diode power probe circuit. (b) Diode quadratic behavior, where the $x$-axis is the corresponding power.

So we can deduce from Eq. (2.8) that, after filtering, which is achieved by the capacitor at the output of the circuit shown in Fig. 2.9(a), the final current under DC operation will be

$$I_D = I_{D0} + \frac{K_2}{2} A^2 \tag{2.9}$$

If the bias current $I_{D0}$ is small or can be eliminated, then the RF power transformed into DC voltage is proportional to the square of the RF input voltage, $A^2$, which is the principle of this type of power probe.

If a modulated signal is used as the input signal, then the mathematics that lead to the DC voltage are much more complex. The reader is referred to [5] for more information.

Modern power probes that are based on diodes are fabricated using GaAs Schottky diodes with a process called planar doped barrier (PDB) technology, which provides improved performance over traditional diode technologies. Modern diode power probes also outperform thermocouples. More information can be gathered from [2].

## 2.2.4 Power-meter architecture

In the previous sections we have seen how power can be measured and how to design and build a power sensor that is based on either a diode or a thermocouple. In both cases, the probe will generate a very small DC voltage, and thus some signal processing should be done prior to the real measurement in the power-meter instrumentation.

The basic power-meter configuration includes the power sensor, which is essentially the power probe, and the power meter itself, where the calculations are done, Fig. 2.10.

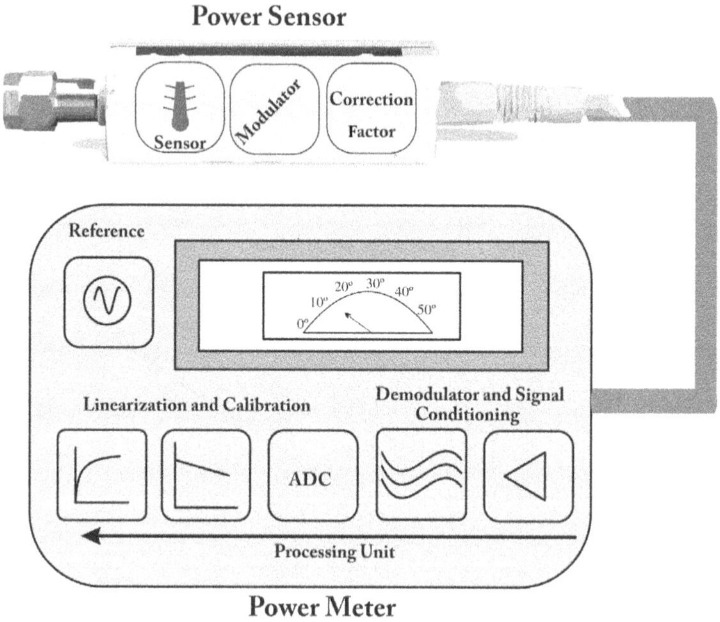


**Figure 2.10** Power-meter block architecture, including a power sensor and the power-meter reading display.

This power meter is most of the time a processor with the capability to measure the DC voltage, apply the calibration corrections (see Section 2.2.5), and then convert it to a power value, shown in watts or decibels on its display.

On looking in more detail at Fig. 2.10, we can clearly identify the power probe, including in this case the power sensor and a signal modulator, which is used for transmission of the DC-value information over a cable, with a high SNR. One possibility is to use a chopper modulator.

The second box that can be seen in Fig. 2.10 is the power meter itself, which consists of amplifiers, a chopper demodulator, and a voltmeter that actually will measure the DC value and then will feed this information to a processor in order to display the correct information. These final stages are implemented by a DSP core in modern power meters.

Finally, a reference oscillator is also seen in the system. This is used for calibration purposes, as will be explained in the next section.

## 2.2.5 Power-meter sources of error

Measurement errors are a cause of incorrect measurements. These errors can be accounted for and sometimes eliminated by using calibration procedures. Nevertheless, there are some uncertainties that cannot be eliminated.

Uncertainty can be defined as "The estimated amount or percentage by which an observed or calculated value may differ from the true value." Thus in our case uncertainty is the amount of measurement error we could expect in percentage terms (%).

In power meters, uncertainty can arise from **power-reference uncertainty**. Since we are using a power reference for our calibration, namely the reference oscillator described above, we should be aware that uncertainty can also arise from errors in the power reference. Reference generators contribute typically an uncertainty of 1% over one year.

Instrument uncertainty can also arise from nonlinearities, attenuator errors, amplifier gain uncertainties, and errors due to the individual components of the instrument itself. Typical values are around 0.5%.

As was seen above, the probe head can behave nonlinearly, especially in diode power probes. In this case tests should be carried out prior to measuring. The tests should include a sweep of power at the input of the probe and then evaluating the plot of output power versus input power, which is a must in order to guarantee a correct calibration of the power probe. This type of error can actually be removed in the case of continuous-wave (CW) signals. By connecting a frequency-swept generator at the input of the probe, and comparing the measured value with the expected value, i.e., the power set by the generator, any problems can be detected. For more complex signals, research to reduce this type of error is under way [5].

Another important source of uncertainty is the mismatch error that arises when measuring a DUT with a power meter that is not matched to the power probe impedance. The error due to the standing wave appearing on this connection will degrade the accuracy of power measurement.

Consider that the probe has a good VSWR, e.g., VSWR= 1.15:1, and that the DUT is significantly mismatched, e.g., VSWR = 2:1. In this case, the overall error will be

$$\begin{aligned} e_{\text{VSWR}} &= 1 - (1 \pm \rho_{\text{DUT}}\rho_{\text{Probe}})^2 \\ &= 1 - [1 \pm (0.333)(0.069)]^2 \\ &\simeq \pm 4.6\% \end{aligned} \tag{2.10}$$

So, in this example, the error is near 0.2 dB in the power measurement. This type of error could be minimized if a match were forced, for instance by using a good-quality, e.g., low-VSWR, attenuator between the DUT and the power sensor, but then the attenuator should be correctly characterized.

Moreover, when measuring the power arising from a DUT, the main objective is to capture all the power going in the probe, called the incident probe power, Fig. 2.11. The incident wave to the power meter is, however, not completely accounted for, since part of the energy is also lost in the instrumentation itself, e.g., loss of heat through the power-sensor package. For this reason, a so-called calibration factor is normally used to account for this error.

The calibration factor is thus nothing other than

$$\begin{aligned} \text{CF} &= \frac{P_{\text{measured}}}{P_{\text{incident}}} \times 100\% \\ &= K_{\text{b}} = \eta_{\text{e}} \frac{P_{\text{gi}}}{P_{\text{i}}} \end{aligned} \tag{2.11}$$

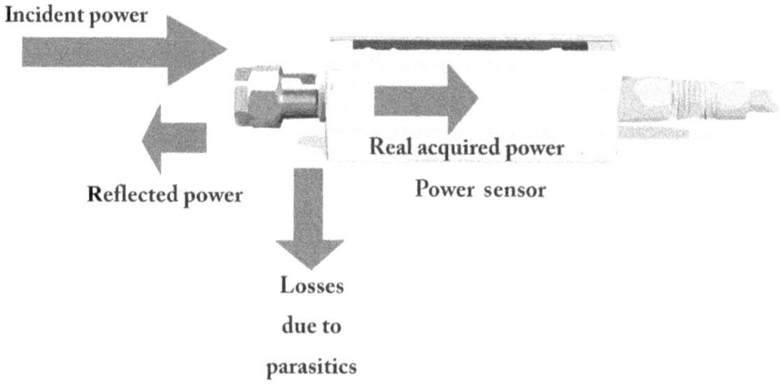


**Figure 2.11** Power-meter calibration needs, where each source of power losses is identified.

where $\eta_e$ is the effective efficiency, $P_{gi}$ is the incident power measured by the power meter, and $P_i$ is the known incident power (the value that should be measured). This frequency-dependent calibration factor is actually unique to each power-meter instrument, and must be measured at the manufacturing stage.

This leads us to the overall uncertainty contributions for a power probe:

$$\begin{aligned} \text{Reference Uncertainty} &= 1\% \\ \text{Instrument Uncertainty} &= 0.5\% \\ \text{Mismatch Uncertainty} &= 4.6\% \\ \text{Calibration Factor Uncertainty} &= 3\% \end{aligned} \tag{2.12}$$

It should be stated that these values are merely indicative. So the worst-case uncertainty would be

$$\text{Uncertainty} = 1\% + 0.5\% + 4.6\% + 3\% = 9.1\%$$

$$\text{Uncertainty (dB)} = \begin{cases} 10\ \log(1 + 0.091) = +0.38\ \text{dB} \\ 10\ \log(1 - 0.091) = -0.41\ \text{dB} \end{cases}$$

The results presented above are obviously the worst-case scenario, and we can expect better results from real measurements.

### 2.2.6 Calibration of the power meter

Power meters should also be calibrated, in order to guarantee the accuracy of the results. By calibration we mean here the act of checking or adjusting the accuracy of the measuring instrument, normally by comparing it with a predefined standard. The need for calibration is mainly due to the fact that the operation of power sensors is dependent on the environmental temperature. For this reason, most power meters, especially those based on a thermocouple sensor, have embedded within them an RF reference oscillator that will allow their correct calibration in situ. On the other hand, the so-called "calibration factors" are already determined and recorded during the power-meter manufacturing process. These are mainly related to the imperfectness of the power meter; see Section 2.2.5.

The main objective of the in-situ calibration is to guarantee that a correct measurement is made, considering several problems that can affect the power meter, such as mismatch of the input probe, parasitic losses, and thermal-error mismatches, that is, incorrect thermal heating due to thermal constraints, Fig. 2.11.

There is some discussion in the R&D literature about the fact that this single-tone carrier calibration is not valid for diode power probes when the aim is to measure highly complex modulated signals. In such cases, the nonlinear dynamics in the diode probe are more complicated, and therefore measurement errors increase. More information can be gathered from [5].

## 2.3 Spectrum analyzers

In the previous section we analyzed and studied power meters. A power meter allows us to measure power over a large bandwidth, but traditional items of equipment do not distinguish at which frequency we are actually measuring the power. Most of the time the signal to be measured is not a pure single-tone CW signal, but may contain harmonics or even be a modulated signal. In this case, the measured power is the total power over the bandwidth of operation of the probe.

Nevertheless, in wireless communications the identification of the power spectral density (PSD), that is, the way the power/energy spans the whole spectrum, is a fundamental need. This is especially true in systems that use the spectrum for improving communications, for instance FDM (frequency-division multiplexing), OFDM, etc. Thus an instrument that captures the power over the spectrum is a must for any wireless communications engineer.

In this section we start by understanding the relationship between time and frequency, and then we proceed by studying different spectrum-analyzer architectures and corresponding figures of merit.

### 2.3.1 The spectrum

Signals are inherently described in the time domain, since that is the natural way of thinking about a process that generates a waveform that varies with time. Actually, a signal can be anything that can be described as a variation over time, $x(t) = f(t)$, where $x(t)$ is the signal itself and $f(t)$ is the function that describes the evolution of the signal over time.

In wireless communications $f(t)$ normally describes a voltage, a current, or any form of energy, and, as we saw previously in Chapter 1, it can be measured and evaluated from different perspectives.

Figure 2.12 presents a typical CW signal in the time domain. As can be seen, the signal, in this case a voltage, varies along the $x$-axis, which corresponds to time.

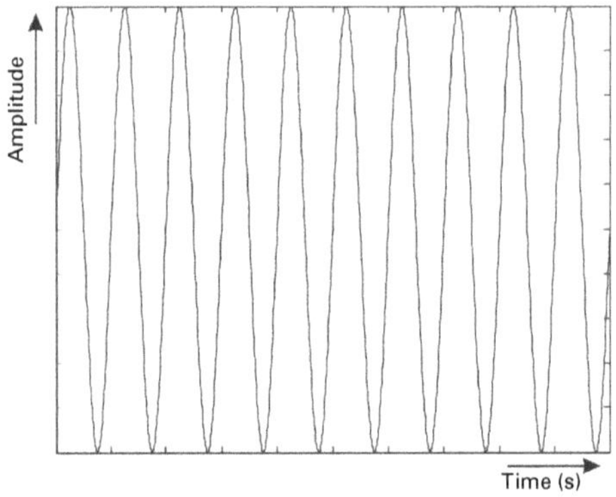


**Figure 2.12** A signal in the time domain, in this case a sine wave.

If the signal is sinusoidal and periodic as in Fig. 2.12, then it can be described as $x(t) = A\cos(\omega t + \theta)$, where $A$ is the signal amplitude, $\omega = 2\pi f$ is the angular frequency, $f$ is the frequency, and $\theta$ is the angle at $t = 0$ s. This is actually the simplest form of signal we can have in a wireless communications scenario.

If the spectrum of the signal $x(t)$ is to be calculated, then the Fourier series can be used.

In the case of periodic signals, we can use the Fourier transform for calculating each coefficient of the Fourier series, which is described as

$$x(t) = A_0 + \sum_{n=1}^{\infty} A_n \sin(n\omega_0 t) + \sum_{n=1}^{\infty} B_n \cos(n\omega_0 t) \tag{2.13}$$

where $A_0$ is the component of the signal at DC, and $A_n$ and $B_n$ are the coefficients of the signal at each frequency component $n\omega_0$, which can be calculated using the Fourier transform as

$$X(\omega) = F[x(t)] = \frac{1}{T}\int_0^T x(t)e^{-j2\pi ft}\,dt \tag{2.14}$$

Then the coefficients of the Fourier series can be calculated from

$$\begin{aligned} A_0 &= \frac{2}{T}\int_{-T/2}^{T/2} x(t)dt \\ A_n &= \frac{2}{T}\int_{-T/2}^{T/2} x(t)\sin(n\omega_0 t)dt \\ B_n &= \frac{2}{T}\int_{-T/2}^{T/2} x(t)\cos(n\omega_0 t)dt \end{aligned} \tag{2.15}$$

Actually, any periodic signal can be decomposed into its harmonic-related coefficients by using Fourier decomposition. For instance, even a square wave can be decomposed by using Fourier coefficients, as illustrated in Fig. 2.13.

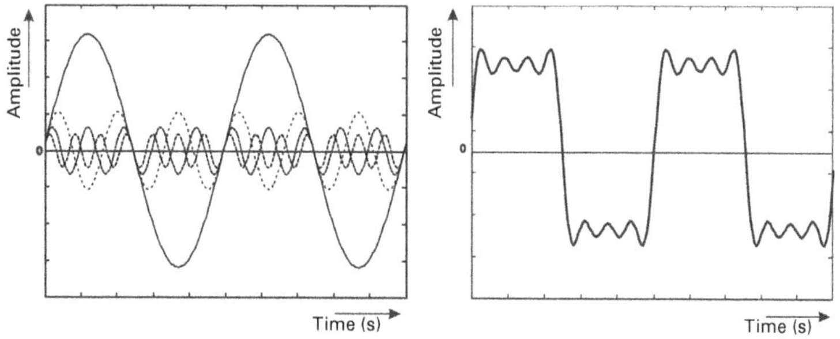


**Figure 2.13** Fourier decomposition of a square wave. In the figure on the left-hand side the first four harmonics are presented, and in the figure on the right-hand side the summation of those harmonics is shown.

For the simplest signal as presented above, $x(t) = A\cos(\omega t)$, the Fourier coefficients are $F[x(t)] = a\delta(f + f_0) + a\delta(f - f_0)$, corresponding to a Dirac function appearing at frequency $f_0$. Remember that the Dirac function is valued 1 at $\delta(0)$ and 0 elsewhere.

This means that each $\delta(f)$ corresponds to a single value at $f_0$ and thus to what is called a spectral line. In the case of a cosine or sine signal, two spectral lines will appear, one at the positive $f_0$ and another one at the negative frequency $-f_0$, but, since negative frequencies are essentially a mathematical artifact, the cosine corresponds in practical use to a single spectral line at $f_0$.

Figure 2.14 presents the spectral decomposition for some known periodic signals.

If the signal is non-periodic, then the Fourier series cannot be applied, and in that case we can apply the Fourier transform in order to gain knowledge of the spectrum occupancy. Actually, the spectrum in the case of non-periodic signals is continuous, and thus spans the whole spectrum, depending on the signal bandwidth.

In this case, spectrum analyzers, as we shall see in the next section, will present the spectrum corresponding to a single time window, and it is considered that it is that time window which should be converted to the spectral domain.

Figure 2.15 presents some non-periodic signal spectra.

Noise is actually a non-periodic signal that should be carefully accounted for, since, as we will see in Section 2.3.2, it can corrupt most of the measured values.

## 2.3.2 Spectrum-analyzer architectures

As we have seen, the spectrum is a mathematical quantity that can be calculated by using Fourier series, but it is very useful for wireless communication signals, since it allows one to decompose a time-domain variation into a summation of spectral lines.

One simple way to build an instrument to measure the spectrum is therefore by implementing the Fourier-series formulation in a hardware platform. By studying Eq. (2.13), it can be deduced that what is being done in that calculation is nothing more than evaluating the signal which appears at each harmonic frequency. So, if we can manage to slice the spectrum into several pieces, and then measure the power in each of those slices, we will have the description and the measurement of the power in

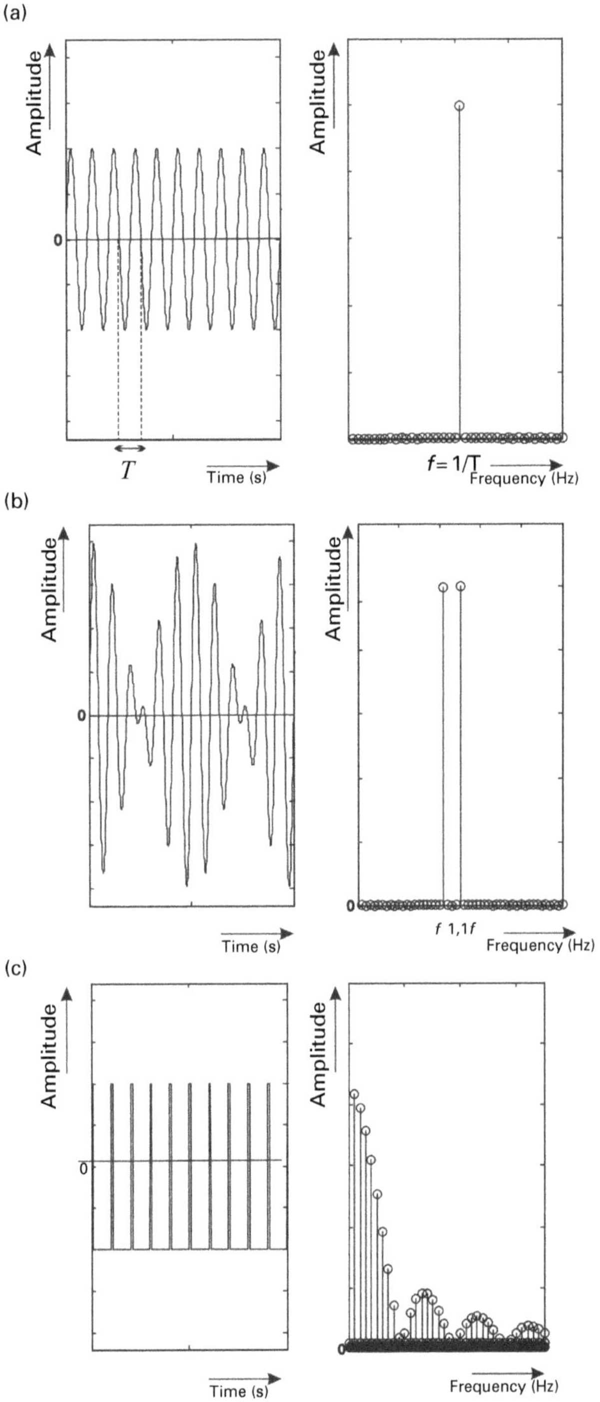


**Figure 2.14** Spectral decomposition of several periodic signals. (a) Spectral decomposition of a single sinusoidal signal. (b) Spectral decomposition of two sinusoids. (c) Spectral decomposition of a square impulse. Only the positive part of the spectrum is shown.

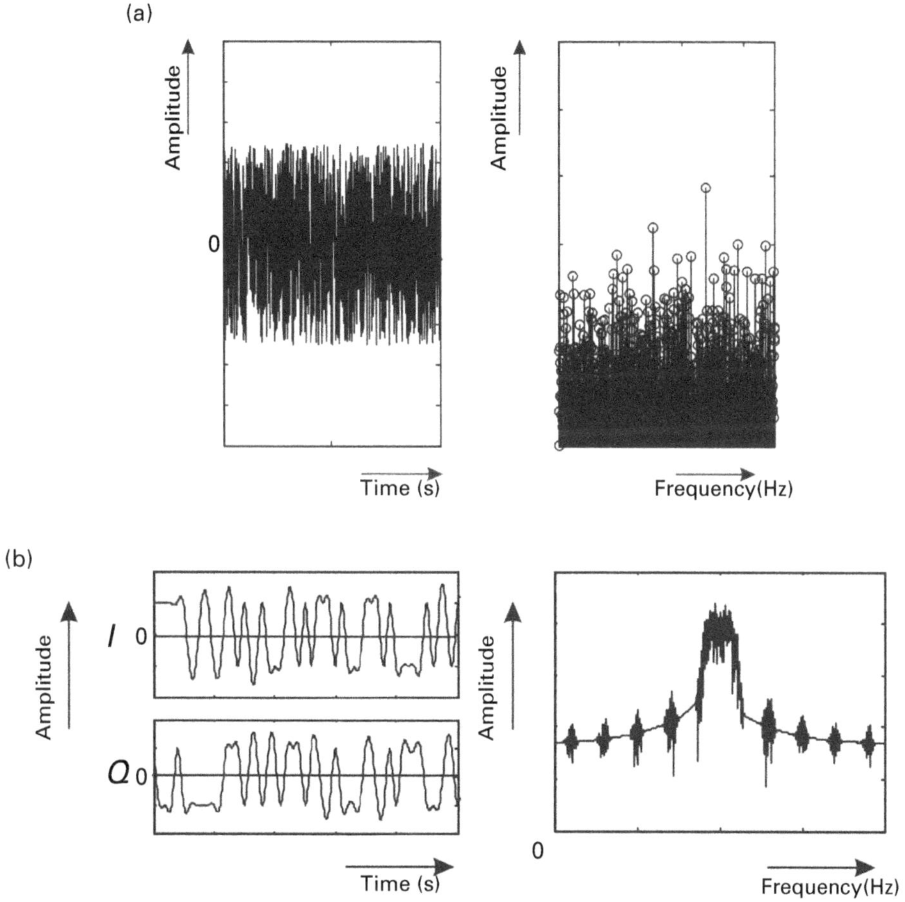


**Figure 2.15** Spectra of non-periodic signals. (a) Spectral decomposition of Gaussian random noise. (b) Spectral decomposition of a real $I/Q$ telecommunication signal.

each slice. If those slices are infinitesimally narrow, a good resolution of the spectrum could be identified. This type of spectrum analyzer is as simple to use as a bank of filters that cut the spectrum into several slices, followed by a power meter for each filter. The power meter could be similar to the one discussed in Section 2.2, since we will measure the power in each filter bandwidth. Figure 2.16 presents this basic concept.

Nevertheless, this implementation is not feasible, since the filters do not have infinitesimally narrow bandwidth, and, even if we decided to sample the spectrum with a small bandwidth, called the resolution bandwidth (RBW) later on, for wireless communications with signals appearing at several GHz, the number of filters would be huge.

Thus a better implementation could be the direct usage of Eq. (2.13), where each input signal is first multiplied by a sine and cosine at the frequency at which we would like to measure the power, and then this quantity is "integrated." This "integration" procedure corresponds to applying a low-pass filter. What this configuration actually does is to sample each spectral component frequency by frequency. The resolution in this case is dependent only on the bandwidth of the low-pass filter, as shown in Fig. 2.17.

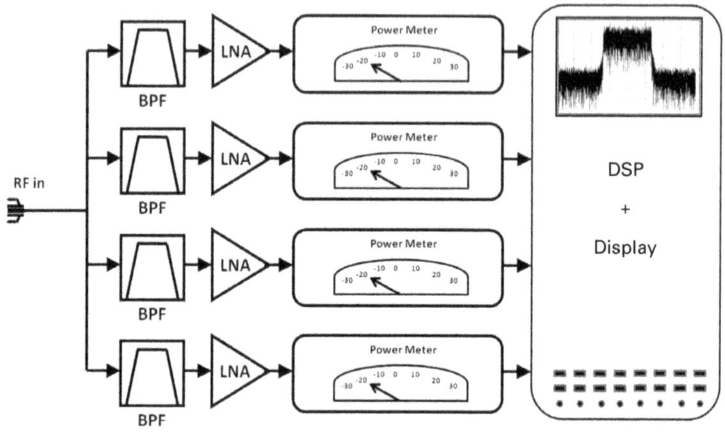


**Figure 2.16** A basic and simple architecture based on power meters and filters.

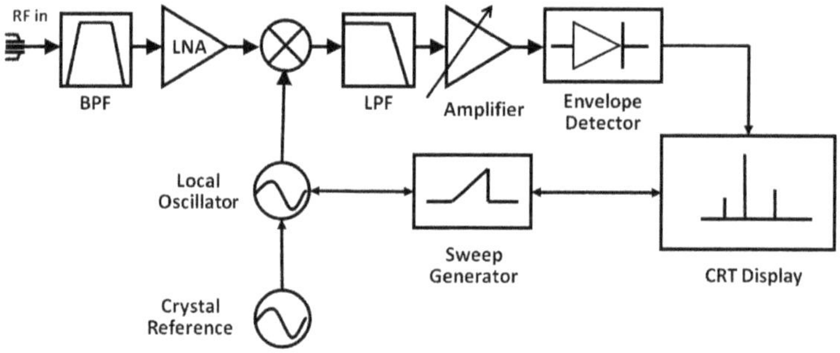


**Figure 2.17** Homodyne configuration.

Looking closely at Fig. 2.17, we see that this is nothing more than a homodyne configuration. Owing to technological limitations in the past, the proposed architecture for most spectrum analyzers is not actually a homodyne configuration, but a heterodyne configuration [6].

The heterodyne configuration multiplies the input signal by a sine wave, but at a frequency slightly different from the frequency of the input signal. The output converted signal, normally at a frequency that is much more reduced than the input signal (but this changes from configuration to configuration, as we will see), is filtered by a bandpass filter, and further evaluated using an envelope detector, which is equivalent to the diode power probe meter of Section 2.2.3. In this case the need for several filters is completely eliminated, since the intermediate-frequency (IF) filter can have a central frequency and a bandwidth that is fixed. The spectrum is thus measured by sweeping the local oscillator frequency and sampling one spectral component at each frequency step. Figure 2.18 presents this configuration.

On looking at Fig. 2.18 we can see several sub-system blocks: an input attenuator, a mixer combined with an oscillator, an IF filter, a logarithmic amplifier, an envelope

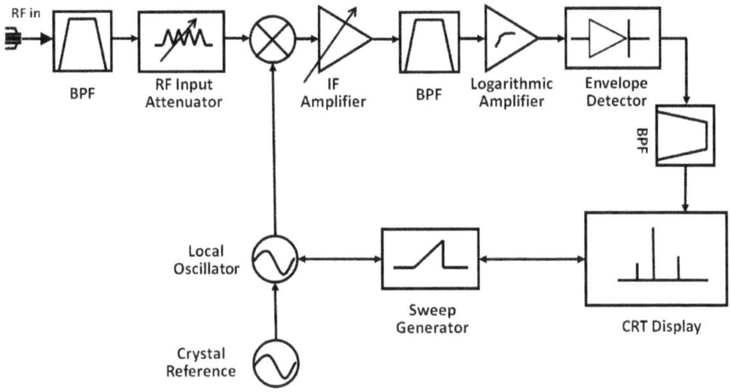


**Figure 2.18** Superheterodyne configuration.

detector, and finally a video bandwidth filter. Let us explain each of these blocks individually.

### 2.3.2.1 The input attenuator

Spectrum analyzers (SAs) are usually designed to present an input impedance of 50 Ω. Therefore some care is needed when the DUT presents a different impedance, since the mismatch between the SA instrument and the DUT can degrade the overall measurement accuracy. As was seen in Section 2.2.5, an attenuator at the input can actually improve this mismatch problem.

The RF input attenuator is normally a step attenuator used for adjusting the signal power level entering the spectrum analyzer. It limits the distortion that will be generated by the spectrum analyzer if the input signal amplitude compresses the input amplifier and/or the mixer. Nevertheless, if an attenuator is used, then the noise factor of the overall system will increase, and the noise floor will be degraded. Thus the attenuation should be selected carefully.

### 2.3.2.2 The mixer and oscillator

The second block includes a mixer and an oscillator. This is the mixing block responsible for the up- or down-conversion of the input signal to the IF. It is responsible for selecting the frequency to be sampled. The oscillator is swept with a ramp curve that controls the oscillator frequency change synchronously with the sweep of the $x$-axis of the display.

The mixing stage converts the input signal by mixing it with the local oscillator frequency. The output signal frequency can be represented by

$$|mf_{\mathrm{LO}} \pm nf_{\mathrm{in}}| = f_{\mathrm{IF}} \tag{2.16}$$

where $m, n = 1, 2, \ldots$, $f_{\mathrm{LO}}$ is the local oscillator frequency, $f_{\mathrm{in}}$ is the input frequency, and $f_{\mathrm{IF}}$ is the intermediate frequency. When only the fundamental frequency is considered in an ideal operation, then $|f_{\mathrm{LO}} \pm f_{\mathrm{in}}| = f_{\mathrm{IF}}$, or the sampled frequency is $f_{\mathrm{in}} = |f_{\mathrm{LO}} \pm f_{\mathrm{IF}}|$.

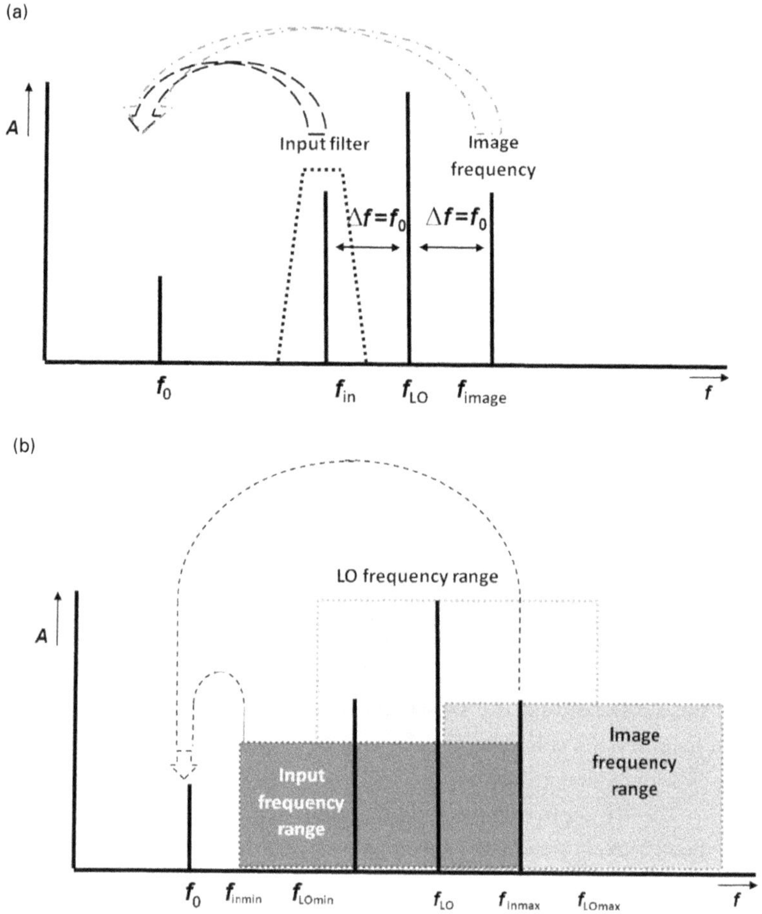


**Figure 2.19** SA mixer operation. (a) Down-conversion of both the input signal component and its image to the same intermediate frequency. (b) The problem with the covered bandwidth.

The problems which superheterodyne receivers suffer from [6] also appear, in this configuration. For instance image-frequencies may appear, and should be eliminated, as illustrated in Fig. 2.19. If the local oscillator has harmonics, then this problem is even worse since several image frequencies can appear at the output IF signal.

To eliminate the image frequency, the input signal image frequencies should be filtered out, but, since the bands can overlap, as seen in Fig. 2.19, the filter should be tunable. This filtering stage can be very complex if the SA has to cover several decades as is usual in recent instruments.

It also should be noticed that most of the SAs available on the market do not allow one to measure DC values, since that can degrade and even damage the input mixers. So they are normally DC blocked, and, if not, an external DC block should be inserted at the input of the SA.

Moreover, the local oscillator inside the SA should have very good characteristics. The local oscillator is usually built using YIG oscillators in analog SAs, and synthesized oscillators in digital ones. In the latter case, the reference oscillator is usually generated by a temperature-controlled crystal oscillator (TCXO). This is the reason why a message saying "OVEN COLD" usually pops up when one switches on old SAs. In analog SAs the local oscillator is swept continuously, but in recent digital SAs, oscillators are normally synthesized and the frequency is swept in small steps, as is visible in Fig. 2.20.

These steps are normally smaller than the RBW in order to reduce errors. It should be noticed that, if the input frequency is not aligned with the IF filter, only a low amount of energy will be measured or no signal will appear, as can be seen in Fig. 2.20. Thus this step method should be carefully designed.

Sometimes, especially for signals with very high frequencies, it may be useful to adopt harmonic mixing, which consists of using higher-order harmonics of the local oscillator to capture higher-frequency signal components. This procedure usually degrades the measurement accuracy due to higher levels of conversion loss. The use of external mixers is also possible if higher frequencies are to be measured.

#### 2.3.2.3 The IF filter

The next block includes a filter combined with a logarithmic amplifier. This is actually the filter which controls the amount of energy which is going to be captured in each sampled frequency, or in other words, it is the filter which calculates each $A_n$ and $B_n$ in Eq. (2.13). The bandwidth of this filter is called the resolution bandwidth (RBW). The IF filter is mainly a bandpass filter that has a variable bandwidth that can be changed by the operator. The combination of the RBW with the sweep time and span (for definitions, see Section 2.3.4.1) are actually the most important characteristics to be selected in a SA, allowing a trade-off among frequency selectivity, measurement speed, and signal-to-noise ratio. As the RBW is narrowed, the selectivity is improved. This can be seen in Fig. 2.21.

#### 2.3.2.4 The envelope detector and video filter

The next block is an envelope detector. It is usually based on a diode detector as presented in Section 2.2.3. This detector measures all the power that is captured by the IF filter and returns the final value for the video filter. The latter is the last block before the display, and it is used to filter out the visual noise. Some modern SAs sample the signal at this stage by using an ADC and then process the video signal in the digital domain, using DSPs or FPGAs.

### 2.3.3 Basic operation of a spectrum analyzer

Let us now see the basic operation of a spectrum analyzer. Considering Fig. 2.22, which represents the heterodyne architecture, the basic operation starts by connecting the signal to be measured to the input attenuator. The value of this attenuator will have to be changed accordingly by the operator, in order to guarantee that the signal will not drive the input blocks into compression, creating distortion, or even damage. As mentioned in Section 2.3.2.1, a higher attenuation value increases the noise level, lowering the SA's dynamic range, and therefore the operator should select the attenuation value with care.

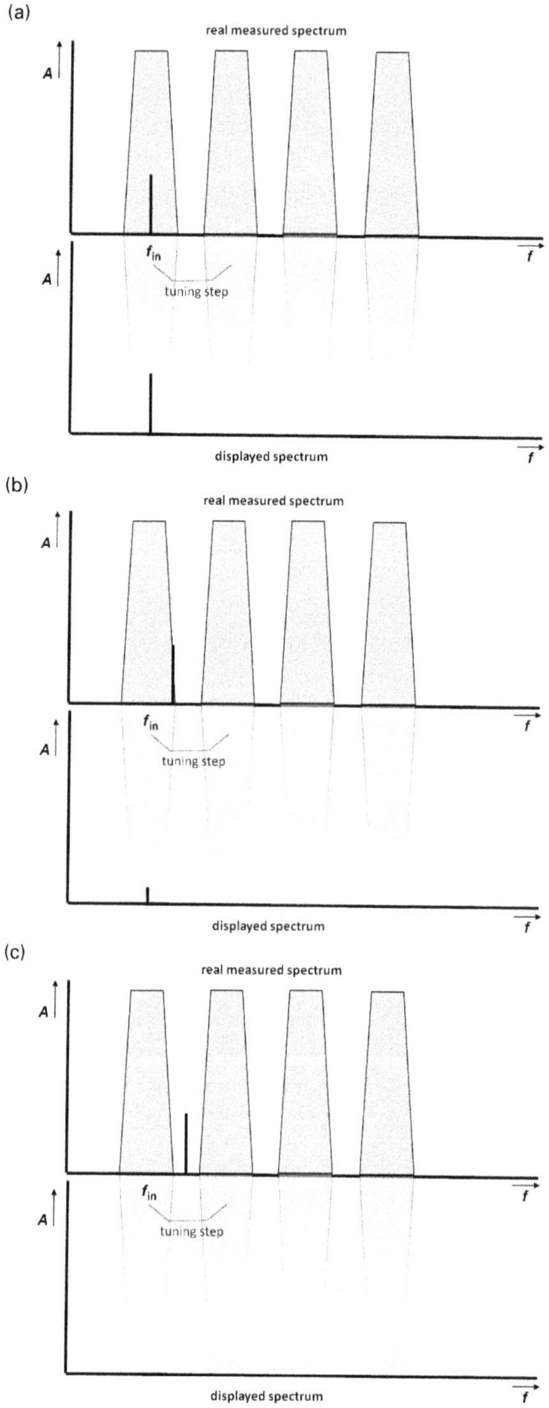


**Figure 2.20** Operation of the stepped oscillator and the IF filter arrangement. (a) The input signal is perfectly inside the IF filter. (b) The input signal is marginally inside the IF filter. (c) The input signal is outside the IF filter.

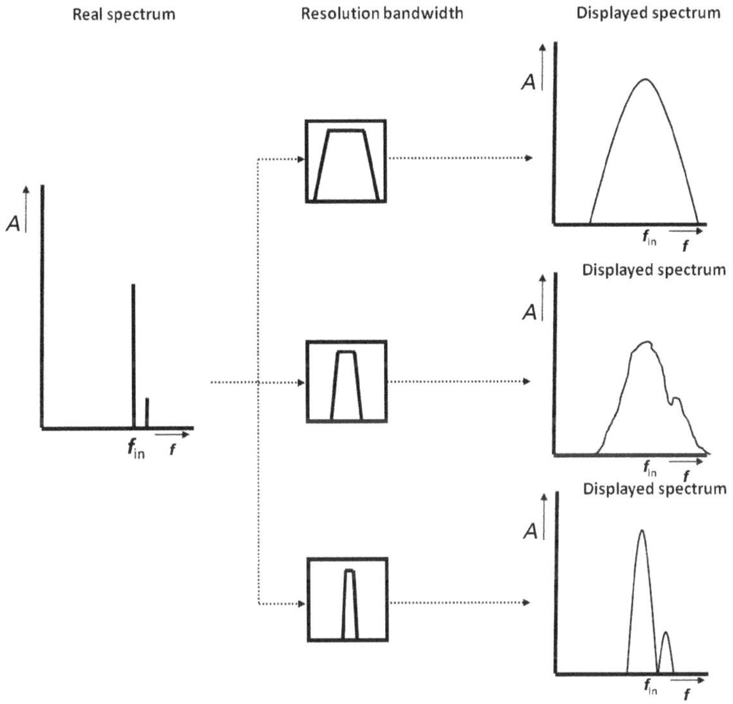


**Figure 2.21** The impact of the resolution bandwidth on measuring two closed signals as seen in the figure on the left-hand side.

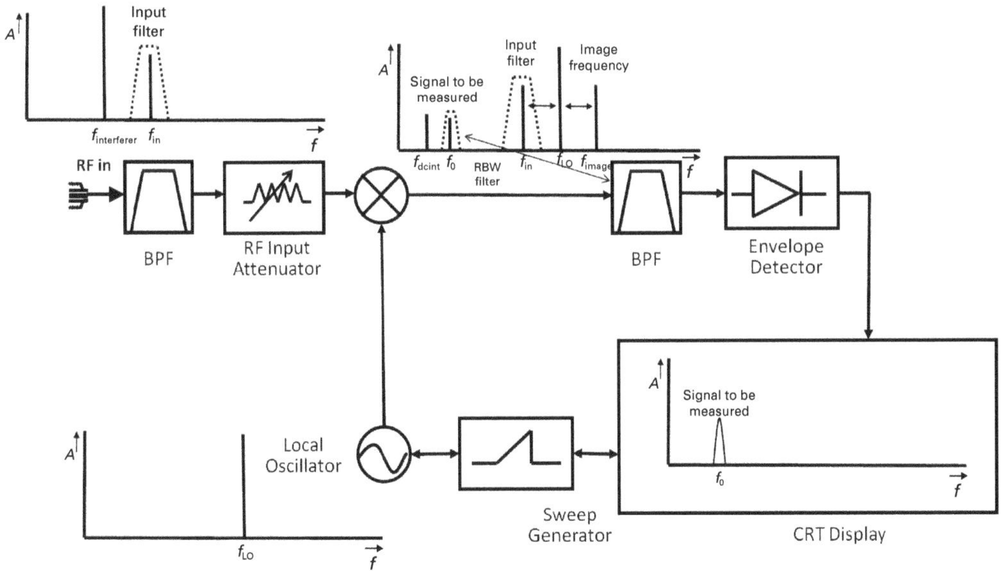


**Figure 2.22** Basic operation of an analog spectrum analyzer.

The signal after traversing the attenuator will be down/up-converted to the IF frequency by the mixer, which will select which frequency will be sampled by using the local oscillator frequency. Despite the fact that most configurations implement a down-conversion scheme, up-conversion can also be considered in certain cases, depending on the bandwidth to be covered and on the selected IF. This IF signal is then filtered out with the IF filter. As mentioned in Section 2.3.2.3, the bandwidth of this filter controls the resolution bandwidth.

The output signal of this IF filter will then be fed to the envelope detector and the power within this band is measured. The signal is then filtered out by a video filter and displayed in a display, either analog or digital.

The frequency to be selected for sampling is determined by a swept generator that controls (in the analog SA) the $x$-axis of the display, which is synchronized with the $y$-axis where the corresponding power arising from the video filter is plotted. In modern displays this is done digitally using a DSP. The speed of the swept generator is actually the sweep time. This should be controlled in order to allow the signal to settle before moving to the next frequency.

Let us now identify the main specifications of a spectrum analyzer.

## 2.3.4 Specifications of a spectrum analyzer

Now that the basic architecture of a spectrum analyzer has been explained and described in detail, let us explain the main parameters of a spectrum analyzer, and how the selected architecture can have an impact on these parameters. In other words, how can we select the correct parameters in order to guarantee that the spectrum analyzer can work in its linear regime and ideally as we predicted?

### 2.3.4.1 Frequency-related parameters

The resolution bandwidth, video bandwidth, sweep time, and span are all related to the spectral behavior, that is, they are related to the dynamics of the instrument. Their values can be selected by tuning and by controlling different sub-blocks within the spectrum analyzer. So let us first explain what each specification means.

### 2.3.4.2 Span

The span is the frequency range over which the spectrum is to be measured. In most spectrum analyzers we can select the frequency range in two ways: either by selecting the start and stop frequency, or by selecting the central frequency and span. In the latter case, span means the overall frequency window that is going to be monitored in the SA (Fig. 2.23).

On looking at Fig. 2.22, we can see that, in order to settle a span value, the SA instrumentation should control the sawtooth-wave generator, so that it covers the frequency range that is selected, and the RF filter should be able to guarantee that the selected frequency window is visible.

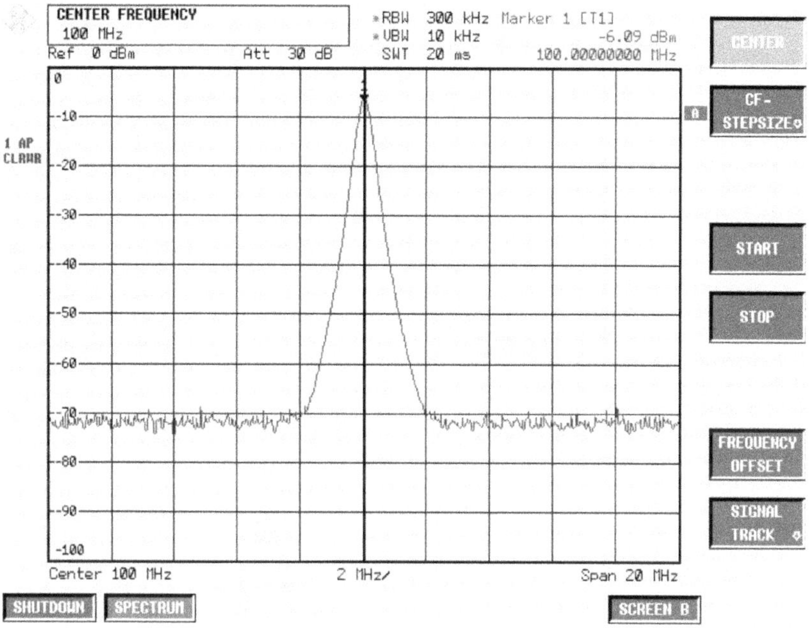


**Figure 2.23** The spectrum-analyzer span and a typical user display. © Rohde & Schwarz.

*Resolution bandwidth*

The resolution bandwidth (RBW) is the frequency resolution that we have on the spectrum-analyzer display (Fig. 2.23). This bandwidth is selected by tuning the IF filter (see also Section 2.3.2.3), and thus by selecting the amount of power that the envelope detector will measure.

Resolution is very important, since it allows the SA to distinguish two different frequencies that can be close together, e.g., see Fig. 2.24, where two different RBWs are selected for evaluating the same signal. When a wide RBW is selected, the SA trace becomes so wide that the upper or lower sidebands are measured as if they were the main signal, masking in that respect nearby signals. Selecting a narrower RBW allows one to increase the resolution, and therefore to further separate the signals, allowing the identification of each one individually. That is what is meant by selectivity.

Selectivity is thus the capability of the SA to distinguish between two or more signals with different frequencies. For instance, some manufacturers define selectivity as the ratio of the 60-dB to 3-dB filter bandwidths.

This filter is also responsible for removing unstable spurious signals, such as the ones arising in the local oscillator, e.g., the residual FM that arises from the contamination of the local oscillator. The IF filter should be able to eliminate this FM noise; if not, the measured signals will appear like the one presented in Fig. 2.25.

The residual FM actually imposes the minimum RBW allowable, and thus the minimum signal-frequency separation that can be obtained with the SA.

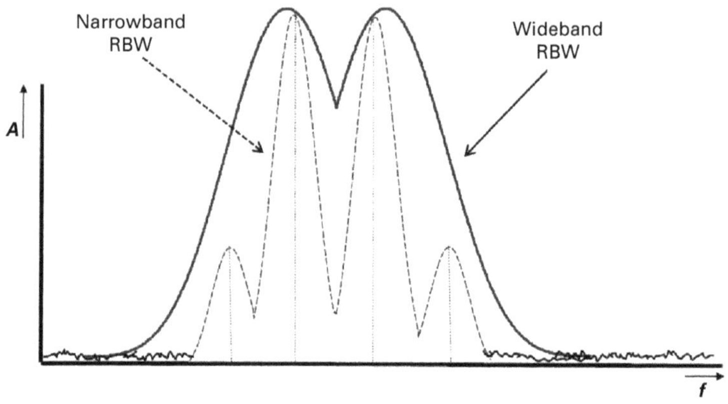


**Figure 2.24** The impact of RBW selection when measuring several close signals in the frequency domain.

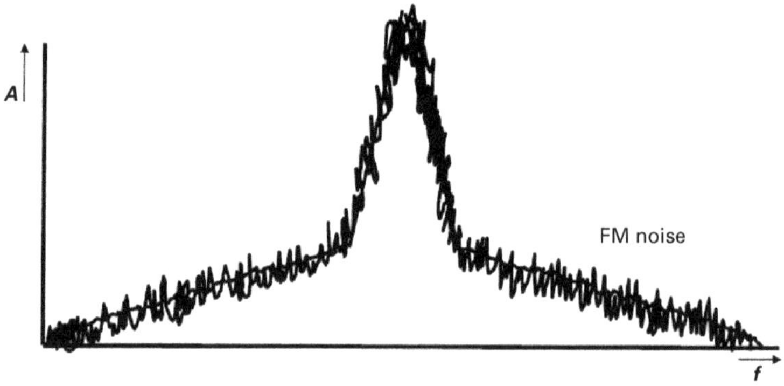


**Figure 2.25** The results displayed by an SA when the LO is severely corrupted by FM noise.

Finally, the oscillator can also contribute instabilities and thus phase noise, which will severely contaminate the measured signal, as can be seen in Fig. 2.26. Phase noise can so severely mask two signals that are close together in frequency that it is no longer possible to distinguish between them.

For instance, consider that we want to measure a signal that is 60 dB down at a 20-kHz offset, using an RBW of 1 kHz. In this case the phase noise that is admissible is $-60\,\text{dBc} - [10\log(1\,\text{kHz}/1\,\text{Hz})] = -60 - 30 = -90$ dBc at 20 kHz. As explained in Chapter 3, this number can be improved by architecture solutions, such as using phase-locked-loop (PLL) oscillators and/or by implementation solutions, such as controlling the temperature of the frequency-generation circuit.

*Sweep time*

By selecting the span of the SA, we are determining the settings of the sawtooth-wave generator that controls the tuning of the local oscillator. We must also select the sweep time of this generator, which is the speed at which the frequencies will be covered. Owing to the fact that the IF (RBW) filter has a settling time, since it is a band-limited

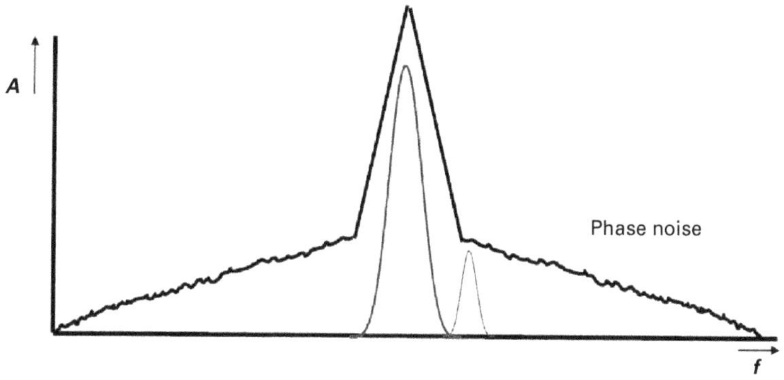


**Figure 2.26** The results displayed by an SA when the LO phase noise is corrupting the measured spectrum.

filter, the selection of the sweep time is crucial to a good selection of tuned frequencies in the SA. If we had no time delay in the filter, then the sweep time could be as small as we want, since each frequency point would be measured instantaneously; unfortunately, due to the limitations of the filter, we should select the maximum allowable sweep time in order to reduce measurement errors. If not, the SA will become uncalibrated. It should also be noticed that narrow RBWs lead to longer sweep times, and wider RBWs to lower sweep times.

The sweep time can be related to the span and RBW by considering that the time during which the signal is maintained within the IF can be calculated as

$$\text{time in IF} = \frac{\text{RBW}}{\text{SPAN/ST}} = \frac{(\text{RBW})(\text{ST})}{\text{SPAN}} \tag{2.17}$$

where SPAN is the span and ST is the sweep time.

Thus, if the rise time of a filter is proportional to its bandwidth, rise time $= k/\text{RBW}$, with $k$ the proportionality constant, then

$$\text{ST} = \frac{k\,\text{SPAN}}{\text{RBW}^2} \tag{2.18}$$

The value of $k$ can vary significantly depending on the filter type, but in modern SAs it can typically be selected as 2.5. It is also common to use several IF filters depending on the selected RBW, and then to use each IF filter only when needed. Modern configurations use digital filters to improve these aspects.

### 2.3.4.3 Amplitude-related characteristics

For the amplitude specifications there are also many parameters that should be accounted for, mainly those regarding the minimum signal level to be measured, and thus associated with the noise floor, and the maximum signal level to be measured, and thus usually associated with the nonlinear distortion introduced by the RF input front end. Let us explain and discuss each of these parameters individually.

*The noise floor*

In a spectrum analyzer one of the most important characteristics to know is the minimum signal level that can actually be measured. This is typically identified by a figure of merit called sensitivity, $S_i$. (The sensitivity is actually different from the selectivity, which is related to the RBW.)

The sensitivity in a receiver is given by

$$S_i = k_B T B F (\mathrm{SNR}) \tag{2.19}$$

where $k_B$ is Boltzmann's constant, $T$ the temperature, $B$ the bandwidth, $F$ the noise figure, which is the amount of noise added by the receiver, and SNR the optimum signal-to-noise ratio to be achieved.

In a spectrum analyzer, the noise floor, $N_{\mathrm{floor}}$, is actually more important than the sensitivity. The noise floor is the power level at which the noise is present, and it can be extracted from the sensitivity as

$$N_{\mathrm{floor}} = k_B T B F \tag{2.20}$$

This equation shows that the noise floor is defined by the thermal noise ($k_B T$), by the bandwidth, in this case the RBW, and by the $F$ of the receiver inside the SA.

As was seen in Section 2.3.4.1, the selection of a correct RBW can reduce the side-lobes, but it can also reduce the amount of noise that is gathered by a factor of

$$\mathrm{noise}_{\mathrm{level}_{\mathrm{change}}} = 10 \log \left( \frac{\mathrm{RBW}_{\mathrm{new}}}{\mathrm{RBW}_{\mathrm{old}}} \right) \tag{2.21}$$

Finally $F$, which is the noise added by the SA, is related to its sub-blocks, but mainly determined by the input low-noise amplifier (LNA). As seen in Chapter 1, this LNA will impose a gain and a minimum $F$ on the input stages of the instrument.

Nevertheless, and in order to reduce the probability of overdriving the input receiver, an RF attenuator is usually included at the input of a SA. This attenuator can significantly increase the amount of noise added by the instrument, since the total $F$ can now be written as

$$F_{\mathrm{total}} = F_{\mathrm{attenuator}} + \frac{F_{\mathrm{amplifier}} - 1}{\mathrm{G}_{\mathrm{attenuator}}} + \cdots \tag{2.22}$$

Thus a 10-dB attenuator will increase the noise floor by at least 10 dB, as can be seen in Fig. 2.27.

The overall noise floor in the SA can then be calculated as $N_{\mathrm{floor}} = 10 \log(k_B T\ \mathrm{RBW}/10^{-3}) + \mathrm{NF}_{\mathrm{SA}} - 2.5$ dB, where the extra $-2.5$ dB is a term included to account for the averaging imposed by the logarithmic amplifier present in conventional SAs.

Video filtering can also be used in this case to reduce the noise, since it will behave as an average filter to the input signal. The video filter is actually a low-pass filter that is included after the logarithmic amplifier (Section 2.3.2.4) and that has no relationship with the RBW sensitivity, rather it has only an averaging effect on the baseband signal. Video filters should have a bandwidth that is equal to or smaller than the RBW filter.

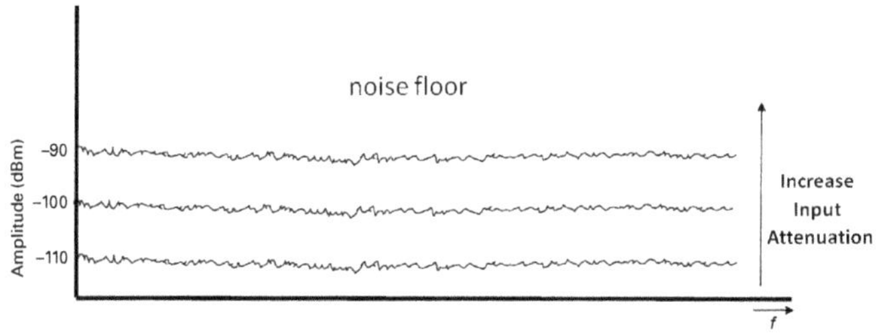


**Figure 2.27** The increase of the SA noise floor with increasing input attenuation.

Typical values span the range from 10 to 100. Owing to the video filter the sweep time should also be updated, using

$$\mathrm{ST} = \frac{k(\mathrm{SPAN})}{(\mathrm{RBW})(\mathrm{VBW})} \tag{2.23}$$

Noise can be further reduced also by averaging the displayed values. This is normally achieved by carrying out several sweeps and then averaging the values. The reduction that can be achieved is given by

$$A_{\mathrm{avg}} = [(n-1)/n]A_{\mathrm{prior}} + (1/n)A_n \tag{2.24}$$

where $A_{\mathrm{avg}}$ is the new average value, $A_{\mathrm{prior}}$ is the average from the prior sweep, $A_n$ is the measured value for the $n$th sweep, and $n$ is the number of sweeps.

Maximum sensitivity can be achieved by selecting minimum RBW, 0 dB input attenuation, and minimum video bandwidth. If extra sensitivity is needed, an external pre-amplifier with improved NF and gain can be added. It should be noticed that, in this case, the pre-amplifier gain should be subtracted from the measured values. Since the input signal to be measured may have a wide spectrum, the pre-amplifier should be accurately characterized over the full bandwidth of the input signal.

*The maximum power*

As follows from Fig. 2.22, it is clear that the front end of the SA is not different from a traditional superheterodyne receiver, with passive, but also active, components [6]. This imposes a linear behavior and also nonlinear behavior in the latter case. For example, the input LNA and mainly the input mixer may create a level of nonlinear distortion that is perfectly comparable to the mechanisms explained in Chapter 1.

So a single tone at the input of the spectrum analyzer will generate harmonics at a certain amount of input power, and a two-tone signal will generate harmonics and intermodulation distortion products. This nonlinear distortion generation will corrupt the measured spectrum and should be avoided or minimized as much as possible,Fig. 2.28.

In the spectrum analyzer we should follow exactly the same procedure as that which we use for other nonlinear components. If we assume that the signals applied impose that the circuits in the SA's front end are working in a mildly nonlinear condition, then the nonlinear behavior can be approximated by a polynomial function as explained in Chapter 1 and in [7]. We can then also define the third-order intercept point, $\mathrm{IP}_3$, and in

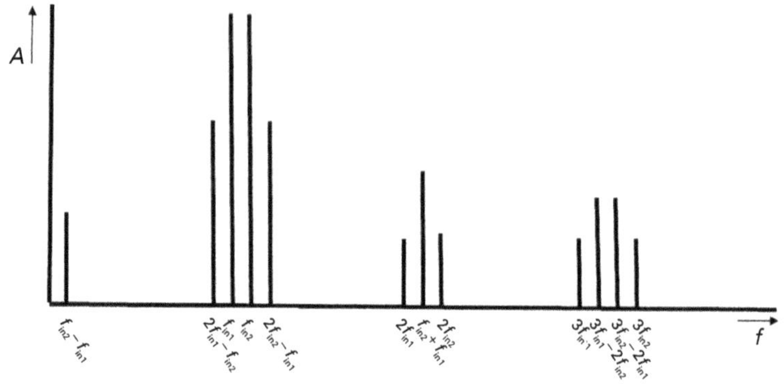


**Figure 2.28** Spectrum-analyzer intermodulation distortion generation at the input stage.

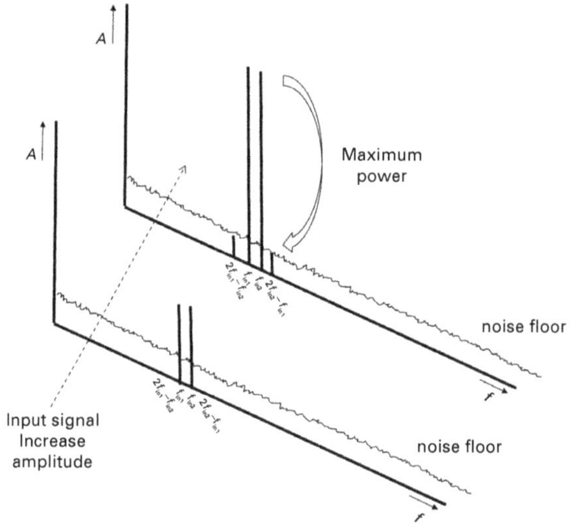


**Figure 2.29** The maximum admissible input power when the intermodulation distortion starts to appear with a level similar to the noise floor.

the mixer case the second-order intercept point, $IP_2$. The maximum value of input power that can be allowed is not determined by the second- or third-order intercept point, but it is the power at which the intermodulation distortion products start to rise above the noise floor. Below this level, we cannot distinguish the intermodulation products from the noise. This is illustrated in Fig. 2.29.

This value can also be calculated from the third-order intercept point $IP_3$. The third-order intermodulation product can be expressed as $IMD_{power} = 3P_{fund} - 2IP_3$. The maximum allowable power is then achieved, when this value approaches the thermal noise floor, namely when $P_{Noise_{floor}} = IMD_{power} = 3P_{fund} - 2IP_3$. A similar calculation is possible for other intercept points.

Figure 2.30 presents the relationship between the input signal power level to be measured and the levels of interference, either noise or distortion, expressed in dBc. For

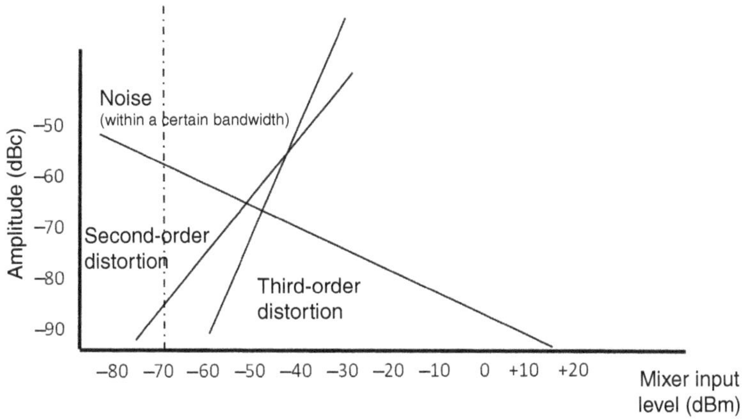


**Figure 2.30** Interference evaluation, mainly noise and intermodulation distortion.

instance, for an input power level of $-70$ dBm at the mixer, the relationship to the noise floor is $-55$ dBc, the second-order intermodulation is at $-85$ dBc and the third-order intermodulation is even further down. As we increase the input power, the noise level moves further away, but the distortion starts to rise, and the difference between the signal level and the distortion level decreases. This will lead us to an optimum input power that corresponds to the intersection of a distortion curve and the noise characteristic. Since the third-order and second-order intermodulation products intercept the noise characteristic at different points, a careful selection considering the most important figure should be made.

Sometimes it is also important to calculate the optimum $\mathrm{IP}_2$ or $\mathrm{IP}_3$ a spectrum analyzer should have prior to acquiring it. This can be done using the previous formulas that relate $\mathrm{IP}_3$ and IMD:

$$\begin{aligned}
\mathrm{IMD}_{\mathrm{power}} &= 3P_{\mathrm{in}} - 2\mathrm{IP3}_{\mathrm{in}} \\
\mathrm{IMR} &= 2(\mathrm{IP3}_{\mathrm{in}} - P_{\mathrm{in}}) \\
\mathrm{IP3}_{\mathrm{in}} &= \frac{\mathrm{IMR}}{2} + P_{\mathrm{in}} \\
\mathrm{IP}N_{\mathrm{in}} &= \frac{\mathrm{IMR}_{N_{\mathrm{order}}}}{N-1} + P_{\mathrm{in}}
\end{aligned} \tag{2.25}$$

For instance, consider that we want to measure a two-tone input signal with which the amplitude of each tone is 20 dBm, and we require that the third-order intermodulation products are at least 60 dB below the input signal level. Thus the $\mathrm{IP3}_{\mathrm{in}}$ of this spectrum analyzer should be $\mathrm{IP3}_{\mathrm{in}} = 60/2 + 20 = 50$ dBm.

As mentioned in Section 2.3.2.1, a technique to reduce the impact of nonlinear distortion in the measurement results is to use an attenuator placed prior to the mixer, which will reduce the power fed to the mixer. This corresponds to an increase in IMR, as can be observed in Fig. 2.30. Nevertheless, we should also be aware that an increase in attenuation will significantly degrade the noise floor, as already explained earlier. So

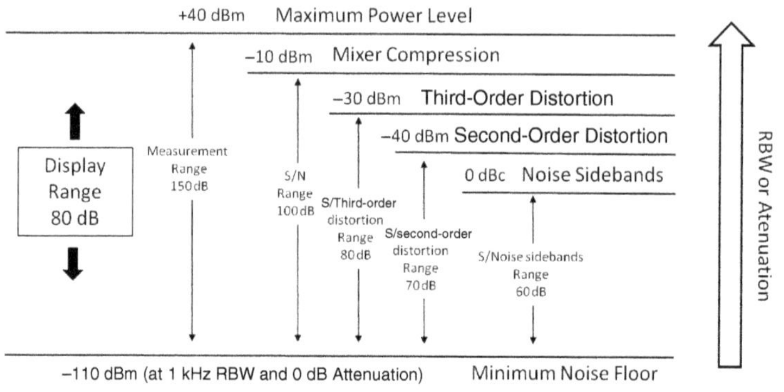


**Figure 2.31** Dynamic-range variation according to different spectrum-analyzer parameters.

a compromise between the attenuation value and the acceptable noise floor should be achieved.

This leads to a practical test that we can use to see whether the nonlinear distortion of the spectrum analyzer is corrupting our measurements. On changing the attenuator level from 0 dB to 10 dB or higher, we should observe an increase in noise on the SA display, and no increase or reduction of the signal being measured. If the signal changes with the attenuation, this means that nonlinear distortion must be present.

*The dynamic range*

The dynamic range is the difference between the maximum power that a circuit or system can handle and the minimum power determined by the noise floor. This definition takes us to the previous explanations in this chapter, where the noise floor and the maximum power were calculated. It should be stated that the instrument's dynamic range, within which we assume that the measured signal is free of interference, and the display dynamic range are normally different quantities, as explained by Fig. 2.31.

The overall dynamic range can be calculated as

$$\begin{aligned} DR_{max} &= \frac{n-1}{n}(IP_{n_{in}} - N_{level}) \\ &= \frac{n-1}{n}(IP_{n_{in}} - 10\log(k_B T) - 10\log(RBW) - NF) \end{aligned} \tag{2.26}$$

where $DR_{max}$ is the maximum dynamic range, $IP_{n_{in}}$ is the intercept point of $n$th order (when the input attenuation is set to 0 dB), $N_{level}$ is the noise level, $n$ is the intermodulation order, RBW is the resolution bandwidth, and NF is the system's noise figure.

The optimum value of the dynamic range can be seen in Fig. 2.32.

## 2.3.5 The accuracy of a spectrum analyzer

When measuring amplitude using any type of instrument, the accuracy should be known, so that we can have confidence in the measured values.

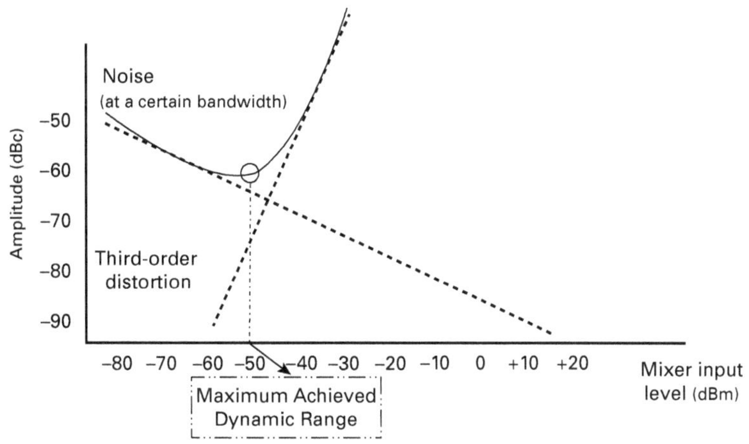


**Figure 2.32** The optimum dynamic-range operation point for a specific application.

Inaccuracy in a spectrum analyzer stems from the non-ideal behavior of several sub-blocks inside the instrument, namely filters, the logarithmic amplifier, the detector, etc. These can have an impact on the amplitude accuracy, but also on the frequency accuracy.

The frequency accuracy depends on the stability of the local oscillator, which can change with temperature, and long-term stability. In a spectrum analyzer the frequency accuracy can actually be calculated using the values presented below:

Aging per year: $1 \times 10^{-7}$
Temperature drift ($+5°C$ to $45°C$) : $1 \times 10^{-7}$
Span error = frequency read from the marker × frequency-reference accuracy
+ 1% of frequency span + 15% of resolution bandwidth
+ 10 Hz "residual error"

by which the frequency accuracy is given by the span error, and the percentages of frequency span and resolution bandwidth are given in the datasheet of the spectrum analyzer.

For example, for a typical SA on the market we have

| | | |
|---|---|---|
| Marker frequency measurement | = 2.4 GHz | |
| Span | = 500 kHz | |
| RBW | = 2 kHz | |
| Span error | = $(2.4 \times 10^9) \times (1 \times 10^{-7})$ | = 240 Hz |
| | 1% 500 kHz | = 5000 Hz |
| | 15% 2 kHz | = 300 Hz |
| | 10 Hz residual | |
| Total | = ±5550 Hz | |

This means that the frequency that we read on the marker on the display is precise with an accuracy of ±5550 Hz.

**Table 2.2** Amplitude accuracy, where $\sigma$ is the variance of each error

| | |
|---|---|
| Maximum uncertainty of amplitude measurement | |
| At 150 MHz, −35 dBm<br>RF attenuation 10 dB, RBW = 1 kHz | <0.2 dB ($\sigma$ = 0.07 dB) |
| Frequency response | |
| ≤50 kHz<br>50 kHz to 3 GHz<br>3 GHz to 7 GHz | <0.5/−1.0 dB<br><0.5 dB ($\sigma$ = 0.17 dB)<br><2.0 dB ($\sigma$ = 0.7 dB) |
| Frequency response with attenuator | |
| 10 MHz to 3 GHz<br>3 GHz to 7 GHz<br>Attenuator<br>Reference-level switching | <1.0 dB ($\sigma$ = 0.33 dB)<br><2.0 dB ($\sigma$ = 0.7 dB)<br><0.2 dB ($\sigma$ = 0.07 dB)<br><0.2 dB ($\sigma$ = 0.07 dB) |
| Display nonlinearity | |
| RBW ≤ 100 kHz<br>0 dB to −70 dB<br>−70 dB to −90 dB<br>RBW ≥ 300 kHz<br>0 dB to −50 dB<br>−50 dB to −70 dB | <br><0.2 dB ($\sigma$ = 0.07 dB)<br><0.5 dB ($\sigma$ = 0.17 dB)<br><br><0.2 dB ($\sigma$ = 0.07 dB)<br><0.5 dB ($\sigma$ = 0.17 dB) |
| Bandwidth-switch uncertainty (RBW = 10 kHz) | |
| 10 Hz to 100 kHz<br>300 kHz to 10 MHz<br>1 Hz to 3 kHz FFT | <0.1 dB ($\sigma$ = 0.03 dB)<br><0.2 dB ($\sigma$ = 0.07 dB)<br><0.2 dB ($\sigma$ = 0.03 dB) |

In terms of amplitude accuracy, Table 2.2 presents some typical values for spectrum analyzers as functions of frequency.

All these values should be added accordingly to calculate the overall amplitude measurement uncertainty, which gives the final error value that we can expect from the measurement. The terms contributing to the overall uncertainty would then be the frequency response (how flat it is over the frequency), attenuator error, IF gain error, linearity error, and bandwidth switching error. The overall uncertainty can then be calculated using the variance of each error contribution, as listed in Table 2.3.

In Table 2.3, $a$ is the maximum systematic error, and $r_s$ and $r_l$ are the mismatch errors.

The total variance of the error is given by $\sigma_{tot} = \sqrt{\sigma_1^2 + \sigma_2^2 + \cdots + \sigma_n^2}$.

Consider a market case with the quantities specified as in Table 2.4.

The overall error variance is then $\sigma_{tot} = 0.39$. For a confidence level of 95%, the value of the specified error expressed in dB is given by $a = k\sigma_{tot}$, where we have $k = \sqrt{2}\,\mathrm{InvErf}\,(\mathrm{CL}/200)$, with CL the confidence level. So for a 95% confidence level $k = 1.96$, and the error is then $e = 1.96 \times 0.39 = 0.76$ dB.

**Table 2.3** Uncertainty in signal analyzers

| Error | Variance |
|---|---|
| Absolute error<br>Frequency response<br>Attenuator error<br>IF gain error<br>Linearity error<br>Bandwidth-switching error | $\sigma^2 = a^2/3$ |
| Bandwidth error | $\sigma^2 = \{10\log(1 + \mathrm{RBW}_{\mathrm{error}}\%/100)\}^2/3$ |
| Mismatch error | $\sigma^2 = \{20\log(1 - r_s r_l)\}^2/2$ |

**Table 2.4** Amplitude accuracy datasheet

| | Specified error | Variance $\sigma_i^2$ |
|---|---|---|
| Absolute error | 0.2 dB | $13.3 \times 10^{-3}$ |
| Frequency response | 0.5 dB | $83.3 \times 10^{-3}$ |
| Attenuation error | 0.2 dB | $13.3 \times 10^{-3}$ |
| IF gain error | 0.2 dB | $13.3 \times 10^{-3}$ |
| Linearity error | 0.2 dB | $13.3 \times 10^{-3}$ |
| Bandwidth-switching error | 0.2 dB | $13.3 \times 10^{-3}$ |
| Mismatch error | | |
| VSWR at spectrum-analyzer input | 1.5 | |
| VSWR at signal-source output | 1.2 | $12.7 \times 10^{-3}$ |

These calculations were done for a single sinusoidal signal. If a modulated signal is to be measured then some extra calculations considering also the root-mean-square (RMS) detector should be done. The reader could refer to [8, 9] for more information.

## 2.4 Vector signal analyzers

Vector signal analyzers (VSAs) are a type of instrumentation that can be considered as an update of the old and traditional spectrum analyzer. In a VSA the signal is subjected to exactly the same process as in an SA, but, before traversing the resolution bandwidth filter, it is sampled using an analog-to-digital converter, and thus it is converted to digital form. The basic architecture is presented in Fig. 2.33.

In this case several problems arise that were not present in the SA. For instance, there is a problem with the dynamic range of the receiver, which in this case is limited by the thermal noise, the maximum power accepted by the mixer, as in the classical SA, but also the number of bits in the ADC is now fundamental, since the signals should be digitized prior to the FFT calculation.

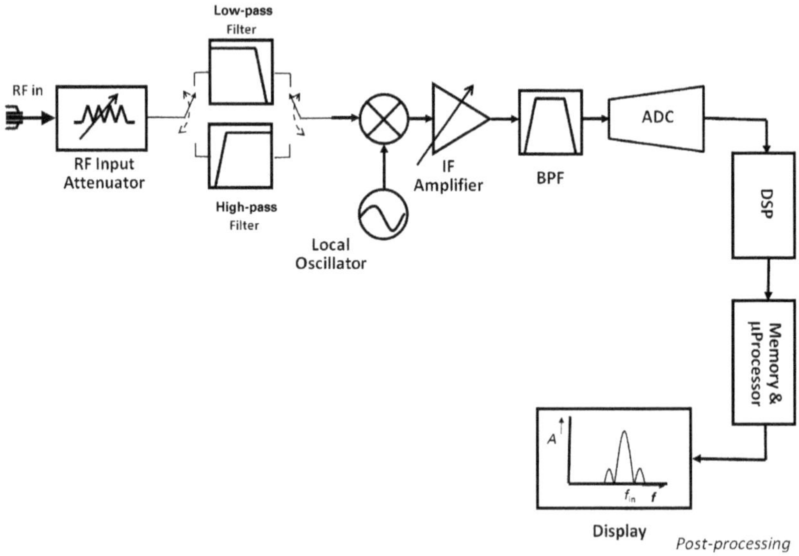


**Figure 2.33** The basic architecture of a vector signal analyzer.

In an ADC, the maximum achievable SINAD ratio can be calculated as SINAD $= 6.02N + 1.76 + \text{OSR}$, where $N$ is the number of bits and OSR is the over-sample ratio. This means that now the main parameter that determines the dynamic range is not necessarily the hardware part, but the sampling and quantization part of the ADC. Moreover, the ADC is controlled by a clock frequency, which defines the sample rate of the ADC, and thus the input signal should fall within the Nyquist band $\omega_{\text{in}} < \omega_{\text{s}}/2$, where $\omega_{\text{in}}$ is the input frequency in radians of the signal, and $\omega_{\text{s}}$ is the sampling frequency in radians.

To better understand this, let us explain the basic operation of the VSA.

## 2.4.1 Basic operation of a vector signal analyzer

On looking at Fig. 2.33, we see that the input signal is down-converted to an IF, filtered out, and then quantized and sampled by the ADC. Sometimes this down-conversion is done using an $I/Q$ demodulator and two parallel ADCs. This procedure is exactly similar to what is used in the SA case, but in this scenario the IF filter will not impose the RBW, but rather will filter out any energy at frequency components appearing above the sampling rate of the ADC in order to reduce any type of aliasing.

Therefore the design of the IF filter is very sensitive, since it has to avoid any aliasing in the digitized version. The filter stage should be designed carefully, since a high amount of filtering provides a certain amount of resolution bandwidth for a certain memory span, while a low amount of filtering limits the span of the measured results (by memory span we mean the number of samples we can save in the instrument's memory). For instance, considering a typical FFT transform, the frequency resolution is given by $\delta\omega = \omega_s/N$, where $N$ is the number of memory points saved in the internal memory. So, for a fixed number of points $N$, the resolution bandwidth can decrease if the IF filter bandwidth decreases, or, for a fixed IF filter bandwidth, the resolution bandwidth can decrease if the number of samples $N$ increases, which will lead to a huge memory requirement.

Later on we will explain this RBW process in depth, but be aware that the number of frequency bins which will give us the frequency resolution $\omega_S/N$ is not the resolution bandwidth of the overall system. Moreover, in this case the maximum frequency to be displayed will be $\omega_S/2$ or smaller due to the FFT's operation procedure.

One possible solution to this dilemma can be the use of digital decimation filters and re-sampling algorithms. Since the explanation of re-sampling algorithms is beyond the scope of this book, the reader is directed to [10] for more information.

Nevertheless, we should retain here that in a decimation filter the sampling rate into the filter is $\omega_S$ and that out of the filter is $\omega_S/n$, where $n$ in this case is the decimation factor. Similarly, the filter output signal bandwidth is $\mathrm{BW}/n$, which is thus a solution for reducing the resolution bandwidth by digital means.

The signal is then digitally processed, as shown in Fig. 2.34.

The first digital process that should be implemented is the use of a digital $I/Q$ demodulator. This will allow the VSA to acquire a digital waveform, and to compare the signal either in amplitude or in phase between the $I$ and $Q$ branches. This is the reason why the VSA is called a "vector signal analyzer," since it measures not only the amplitude of digitally modulated signals, but also the phase, or, in other words, vector information. Note that the vector in this case is related not to the RF part of the signal, but rather to the baseband part of the signal after $I/Q$ demodulation. This is in contrast to the nonlinear network analyzer instrumentation covered in Section 2.7.

This $I/Q$ signal is then filtered out by digital means, and will fill up a memory. After the memory has been filled up, the DSP will apply different mathematical approaches to the data, and the spectrum can be obtained using an FFT. So the spectrum is obtained much as in a traditional SA, but the RBW, span etc., are now chosen by digital means.

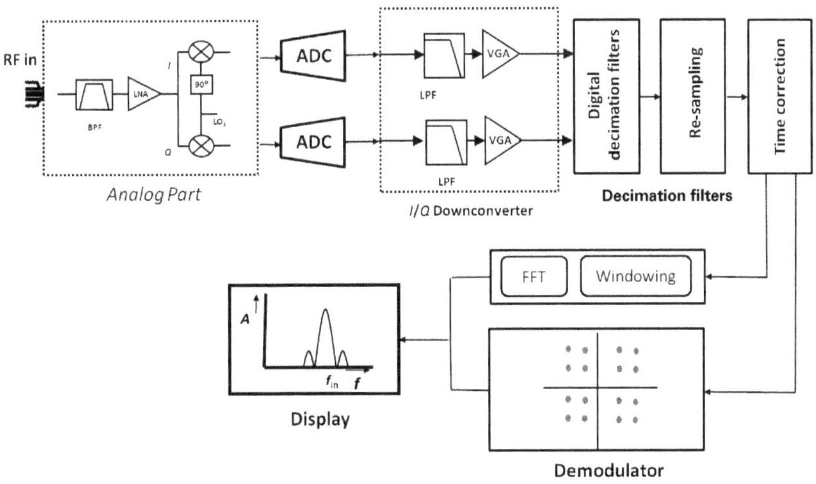


**Figure 2.34** The digital side of the vector signal analyzer.

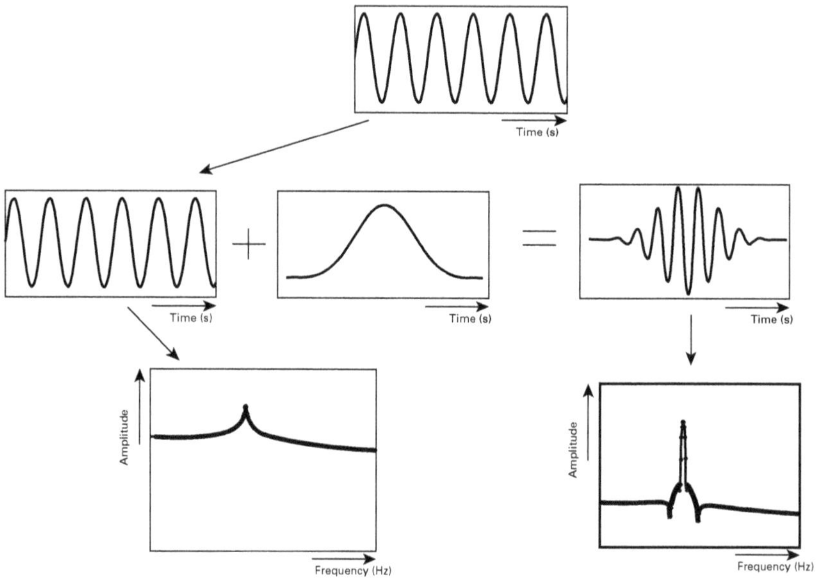


**Figure 2.35** The windowing process of a vector signal analyzer.

**Table 2.5** Window types for spectral evaluation

| Window | Typical application |
|---|---|
| Uniform | Transient evaluation |
| Flat top | High amplitude accuracy |
| Gaussian top | High dynamic range |
| Hanning | general purpose |

Again it should be noted that the maximum span is defined by the IF filter. Moreover, the data is first passed through a window function that will obviate what is called spectral leakage. The window function attenuates the start and end of the time-domain signal, as can be seen in Fig. 2.35.

Typical windows include the ones presented in Table 2.5.

After the application of these windows, the new resolution bandwidth can be calculated as $\text{RBW} = \text{ENBW}/T$, where ENBW is the equivalent noise bandwidth, which corresponds to the bandwidth that a traditional filter would need to have in order to generate the same amount of white noise, and $T$ is the time-record length.

It should also be noted that the real-time bandwidth is not necessarily what is displayed on the VSA screen, since there is a processing time that can delay the display of the FFT. For instance, if we have 800 line measurements using a 2-kHz span, this will require a 400-ms time record, and if the number of line measurements increases then the delay time will be higher.

This fact implies that, on certain occasions, when there is no time to process the FFT, some input data is lost, as can be seen in Fig. 2.36.

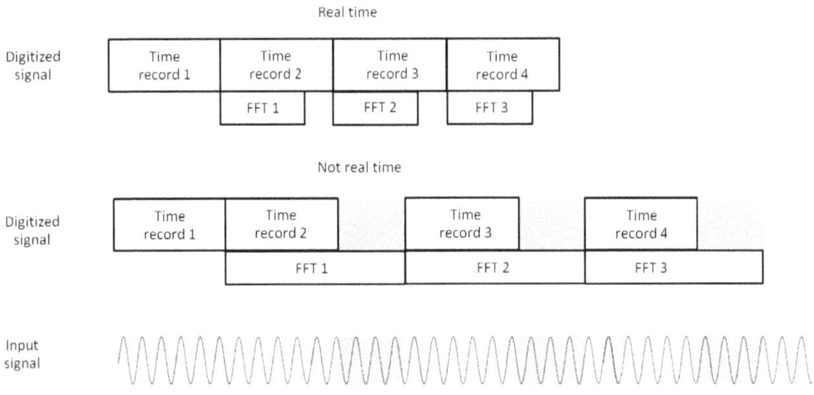


**Figure 2.36** Input signal analysis for a vector signal analyzer.

The data can also follow other data-processing mechanisms, and thus other types of information, such as constellation diagrams, symbol streams over time, demodulated signal information, the error-vector magnitude, etc. can be obtained.

A VSA can also work as a time-domain scope, similarly to a digital oscilloscope, but for the baseband signal only. So the displayed signal will be not the RF signal, but rather its down-converted version. Calibration and uncertainty in a VSA are similar to those presented for the SA, since the analog front end is actually similar.

## 2.5 Real-time signal analyzers

In this section we will now discuss another type of spectrum analyzer, which makes use of the VSA concept and applies it to a powerful FFT analyzing technique. Actually it calculates FFTs with a high-speed processor, allowing one to create what is called a real-time signal analyzer (RTSA).

In a traditional VSA, where the FFT is implemented at the intermediate frequency, the configuration imposes a requirement for a certain amount of time to calculate the Fourier transform, and during that time the input signal is ignored, as illustrated in Fig. 2.37.

This means that a VSA can ignore and can be blind to signals that appear only during certain periods of time. On looking at Fig. 2.37, we see that the first spike on the time-domain waveform is completely ignored, since the processor was calculating the FFT during that time.

In the RTSA concept, the processors are used in parallel, mainly using different ASICS, and so not one but several FFTs are implemented simultaneously. This allows the instrument to acquire the data and to calculate FFTs at the same time, over a common signal, reducing or even eliminating the blind spots that appear with VSAs. This was achieved due to the significant increases in processor speed and memory of which these new instruments can avail themselves. Figure 2.38 presents the parallel- FFT concept.

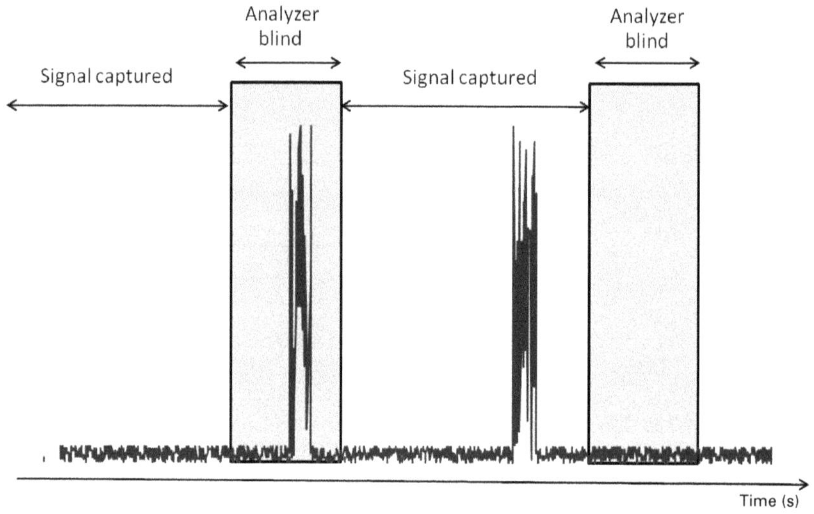


**Figure 2.37** Blind spots for a vector signal analyzer.

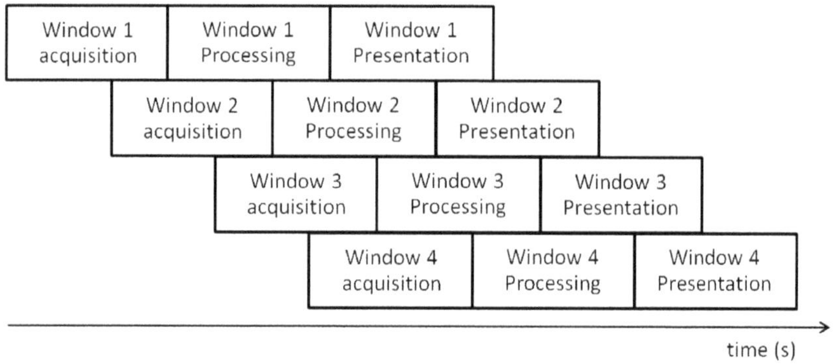


**Figure 2.38** Real-time analyzer FFT processing.

Thus an RTSA is an instrument that allows parallel sampling and FFT calculation. This means that the sampler is working during the calculation of the FFTs executed by different FPGAs. Thus, before a new window is captured, the FFT has already been saved in memory. In fact the FFTs that are implemented in an RTSA are most of the time so-called short-time Fourier transforms. Since it is a fast FFT and the number of acquisition points is limited, typically 1024, the processing time can be shortened.

Another important characteristic that can be seen in Fig. 2.38 is the overlap of the FFT calculation. Since we are implementing FFTs in a limited time window, we should be aware that some errors will exist due to the windowing of the input signal. Thus the overlapping of FFTs, meaning overlapping of captured windows, allows subsequent processing and spectrum reconstruction.

## 2.5.1 The RTSA block diagram

The main blocks of an RTSA are shown in Fig. 2.39. The configuration is similar to the one presented for the VSA, but with the difference that now we capture the full $I/Q$ signal and convert it in a fast way using a short-time Fourier transform (STFT). The STFT is expressed by

$$\mathrm{STFT}\{x(t)\} = X(\tau, \omega) = \int_{-\infty}^{\infty} x(t)w(t-\tau)e^{-j\omega t}\,dt$$

$$\mathrm{STFT}\{x(n)\} = X(m, \omega_m) = \sum_{n=0}^{N-1} x(n)w(n-m)e^{-j\omega_m n} \tag{2.27}$$

where $w(t-\tau)$ or $w(n-m)$ is the window function used for the STFT evaluation.

On the digital side, presented in Fig. 2.40, the main components are the parallel computations of FFTs using a fast FPGA, allowing capture of the signal and its subsequent processing to display the final information.

It should be stressed again that the main property of the RTSA is the capability to calculate the spectrum in a very short time frame, and that is why the instrument is called a real-time signal analyzer.

The RTSA also brings some novel important concepts to spectrum analysis, mainly due to the fact that the RTSA keeps the spectrum's history saved in memory. Thus concepts such as spectrograms, persistence displays, and spectrum triggers allow the user to gain a new vision of spectrum analysis.

Let us now explain these concepts individually.

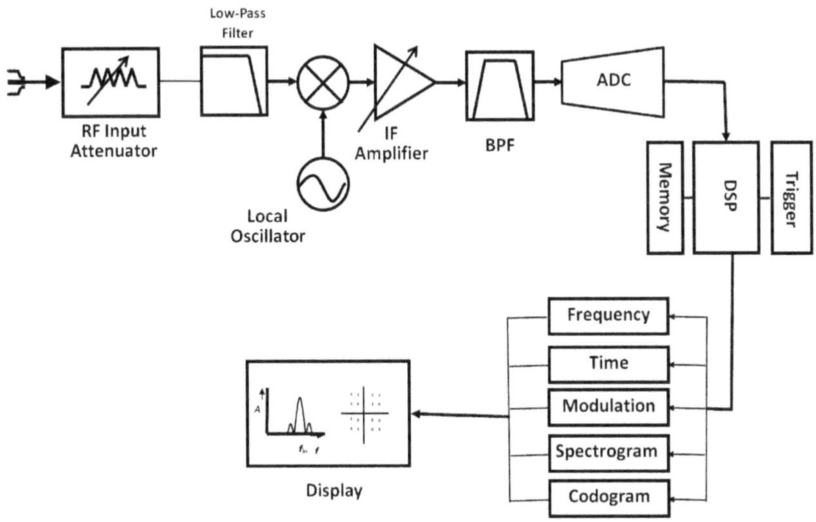


**Figure 2.39** The block diagram of a real-time analyzer.

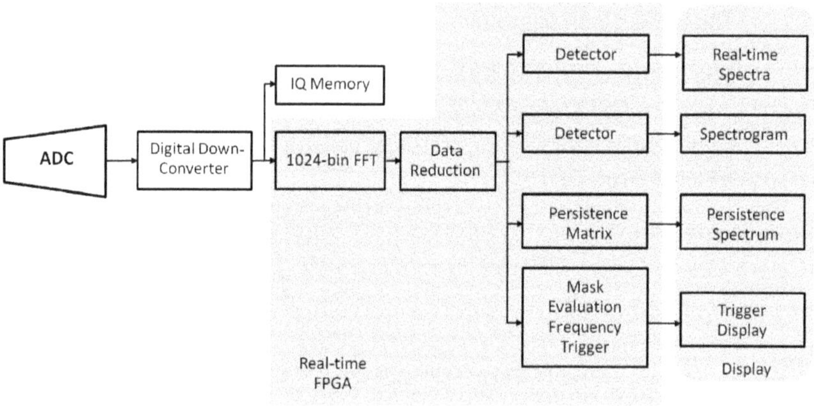


**Figure 2.40** The digital block diagram of a real-time signal analyzer.

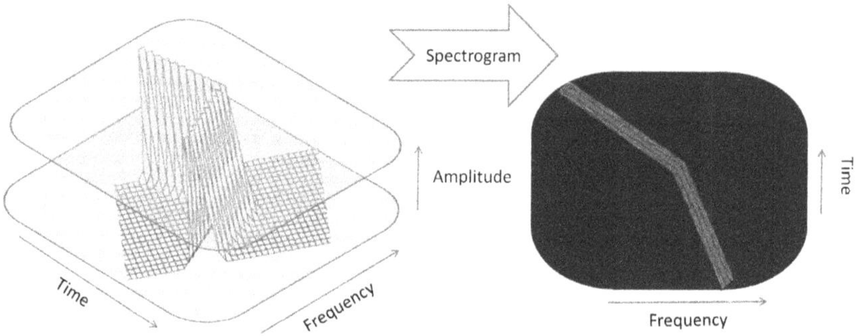


**Figure 2.41** The digital spectrogram of a real-time signal analyzer.

## 2.5.2 The RTSA spectrogram

Since the RTSA has all the individual windows' spectra saved in memory, it can present a diagram in three-dimensional form that visualizes the evolution of the spectra through time. This display is called the spectrogram. A spectrogram represents the spectra, but sliced in time. Figure 2.41 presents this concept.

In the RTSA the spectrogram is usually plotted in two dimensions using a color graph that represents the amplitude of each spectral bin by using a hotter or colder color, namely red or blue, respectively.

In this operation mode the user can actually select the span and the RBW as in previous VSA and SA solutions, but now we should be aware that these two values are related, since the calculation time of the FFT has to be accounted for, even if it is small when compared with that for a traditional VSA. Since in most RTSAs the number of FFT bins is fixed at 1024, to speed up the algorithms, the span/RBW pair is fixed, which means that when the span is reduced the RBW is also reduced and vice versa.

Other parameters can also be tuned, such as the sweep time, which actually tells the instrument how many FFTs it should combine prior to presenting the display, the history

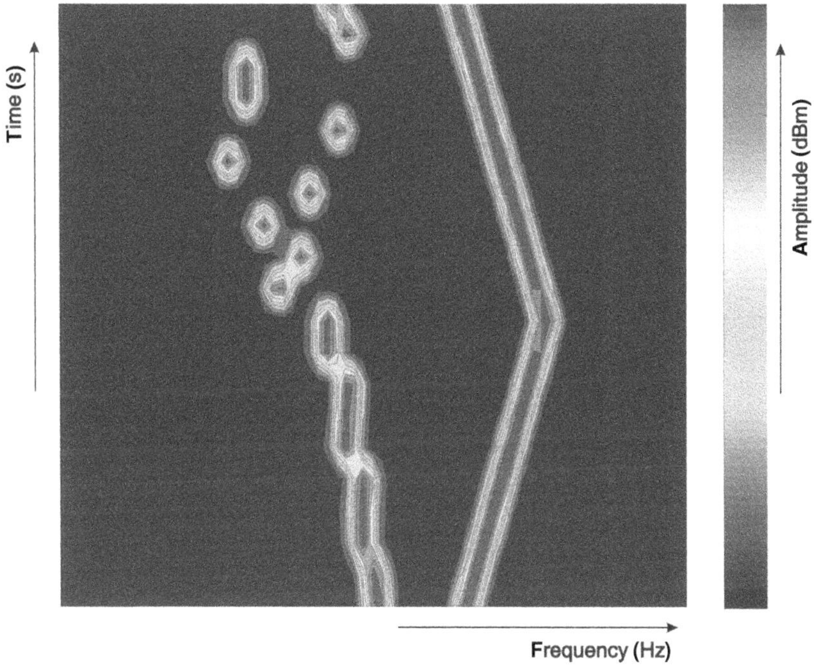


**Figure 2.42** The digital spectrogram of a real-time signal analyzer.

depth, which informs the instrument how many spectra it should plot, and also the color-mapping function, which defines the colors to be used and the minimum and maximum thresholds whence to start painting the spectra. Figure 2.42 presents a real spectrogram in an RTSA.

## 2.5.3 RTSA persistence

Another way to observe real-time spectra is to use persistence maps. Persistence maps actually work as explained in Fig. 2.43, where the spectra will start populating a display graph, and thus will create an image over time that corresponds to the number of times the spectra pass through that specific amplitude point. So, after some time, the graph colors will give one some idea of how the spectra behave in a transient fashion.

The main tuning parameters are the span and RBW. The operation and limits are equivalent to the spectrogram case. Nevertheless, some other parameters can be selected, such as the persistence granularity, which is somewhat similar to the history depth in the spectrogram. This parameter tells the instrument how many points it should count for, or, in other words, the time duration for which the graph is to be plotted. After that time the graph is restarted.

Another parameter is the persistence, which determines the amount of time until a trace has completely faded. This is equivalent to what was seen in old and traditional cathode-ray-tube (CRT) instruments. Some other parameters include the maximum hold

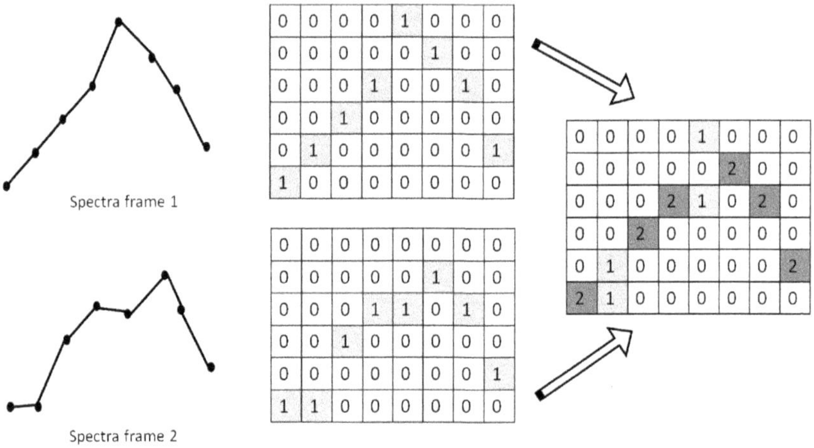


**Figure 2.43** The persistence of a real-time signal analyzer.

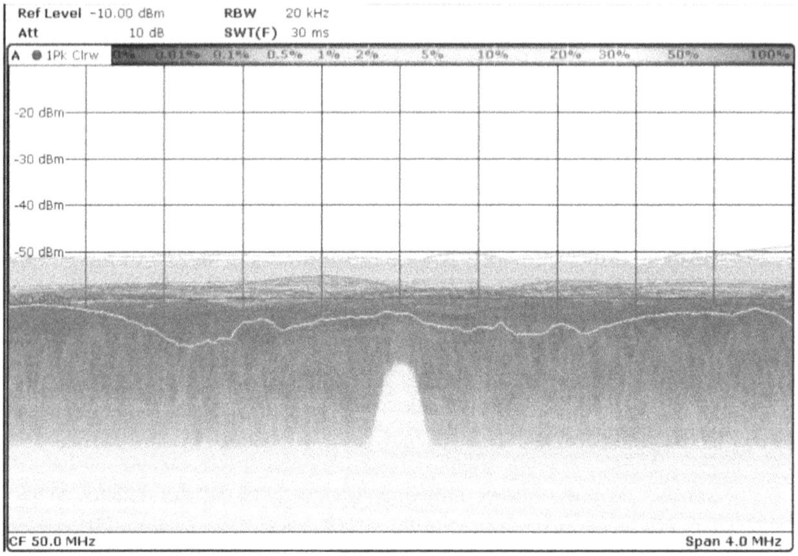


**Figure 2.44** The persistence graph of a real-time signal analyzer. © Tektronix.

intensity, to guarantee that the maxima are observed in the color mapping, similarly to what was seen with the spectrogram. Figure 2.44 presents a real persistence display in an RTSA.

### 2.5.4 The RTSA spectrum trigger

Finally a new improvement in spectral analysis was also introduced by the RTSA, called the spectrum trigger. It means that the signal is measured and saved only when the

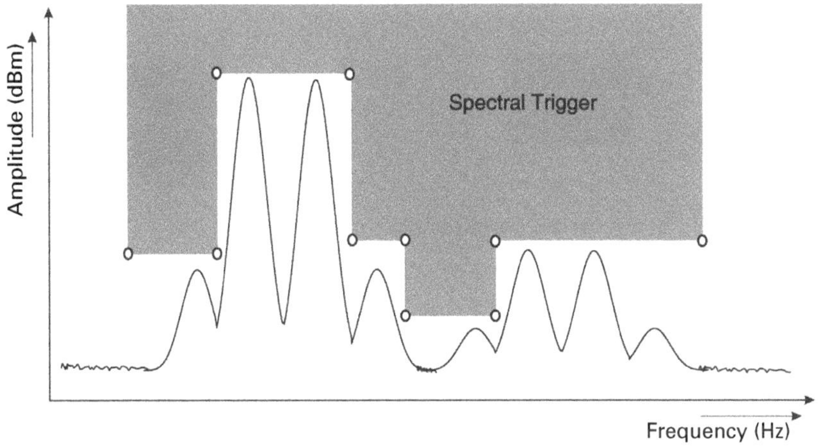


**Figure 2.45** A spectral trigger for a real-time signal analyzer.

spectrum touches a predetermined spectral mask. This measurement is very important and useful for engineers who want to guarantee that a circuit/system does not transmit outside the predetermined mask for which it was designed.

On considering the previous section, we see that we divide the spectra into three vectors, namely frequency, time, and amplitude. So the spectral trigger is nothing more than an indication to start an event. Most of the time the event consists of starting to save the spectral information after a trigger has been activated, and then stopping when the spectral information moves out of the triggering mask, as can be seen in Fig. 2.45.

There are also other ways to achieve spectral triggering, including entering the trigger mask, leaving the trigger mask, etc. Figure 2.45 presents a real spectral triggering process.

## 2.6 Vector network analyzers

A vector network analyzer (VNA) enables the measurement of $S$-parameters, which, as described in Chapter 1, are one of the main important figures of merit to be measured by microwave and wireless engineers. $S$-parameters relate incident and scattered traveling voltage waves at the ports of a microwave circuit. More precisely, $S$-parameters correspond to *ratios* of scattered to incident traveling voltage waves, which is important to keep in mind to understand the calibration process. The expressions for a two-port device are

$$S_{11} = \left.\frac{b_1}{a_1}\right|_{a_2=0}$$
$$S_{12} = \left.\frac{b_1}{a_2}\right|_{a_1=0} \qquad (2.28)$$

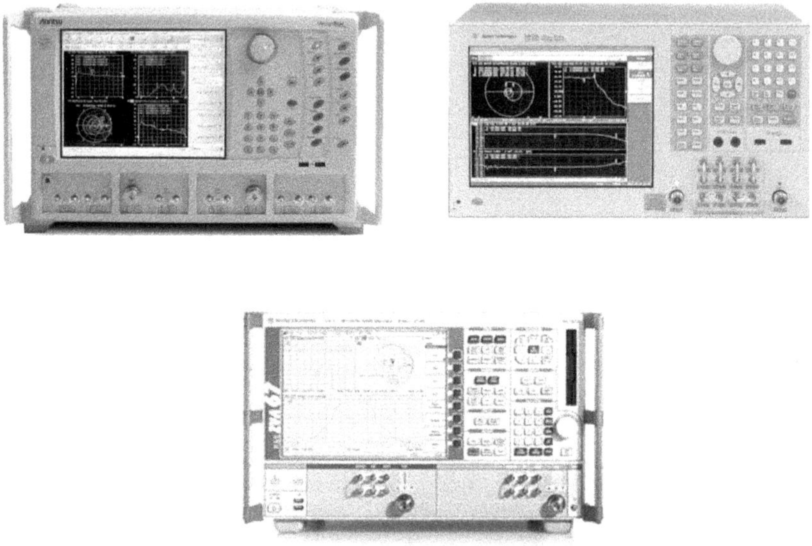

**Figure 2.46** Two-port VNAs: from left to right, Anritsu, Agilent Technologies, and Rohde & Schwarz. © Anritsu, Agilent Technologies, and Rohde & Schwarz.

$$S_{21} = \left.\frac{b_2}{a_1}\right|_{a_2=0}$$

$$S_{22} = \left.\frac{b_2}{a_2}\right|_{a_1=0}$$

with $a_i$ and $b_i$ the incident and scattered traveling voltage waves at port $i$, respectively.

We will first explain the instrument's architecture in Section 2.6.1, followed by the calibration procedure in Section 2.6.2. In the following, the two-port implementation is considered, since this is the most common instrument in daily use. See Fig. 2.46 for practical examples. The architecture and calibration can be straightforwardly extended to multi-port (>2 ports) VNAs.

## 2.6.1 Architecture

The basic architecture of a two-port VNA is presented in Fig. 2.47. The DUT can be either connectorized or on-wafer. The architecture consists of at least one RF signal generator, two directional couplers, and a receiver. These blocks are now discussed in more detail.

### 2.6.1.1 The signal generator

A signal generator is required in order to generate the incident traveling voltage wave. If the VNA has only one RF source, an internal switch takes care of applying the source alternately to port 1 and port 2 of the DUT. The other port is terminated in 50 Ω. Modern

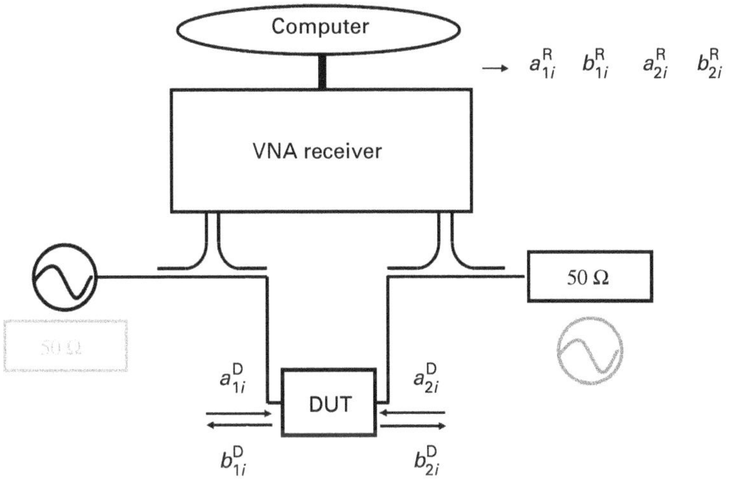


**Figure 2.47** The fundamental blocks of a VNA.

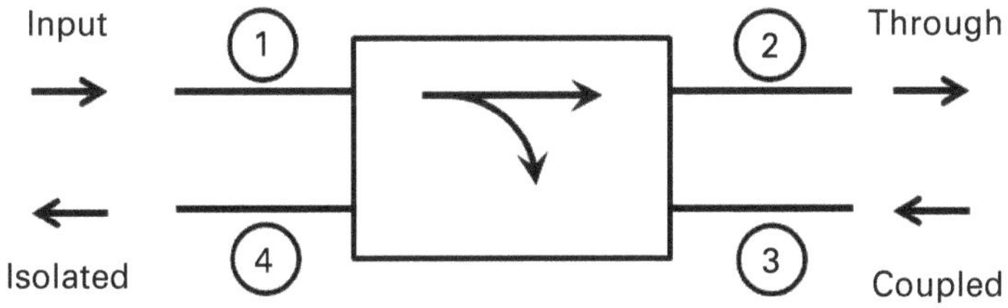


**Figure 2.48** A directional coupler.

network analyzers often have two internal RF generators, one for each port. An in-depth explanation of the internal architecture of an RF signal generator is presented in Section 3.2. Most of these signal generators can be swept in frequency and power. The latter option is of interest to measure the AM–AM and AM–PM characteristics (see Section 1.5.1).

### 2.6.1.2 The directional coupler

At each port a directional coupler is used to separate the incident and scattered traveling voltage waves. A directional coupler is a four-port passive device, as illustrated in Fig. 2.48.

Assume that the directional coupler of Fig. 2.48 is positioned at port 1 of the DUT. A similar analysis can be carried out for port 2 of the DUT. In this case, port 1 of the directional coupler is connected to the RF signal generator, and port 2, also called the "through port," is connected to port 1 of the DUT. A fraction of the power injected at port 1 is directed to port 3, which is also called the "coupled port." This fraction is small in the case of VNA measurements, such that most of the power generated by the

RF signal generator passes on to the DUT via the through port. The coupling is defined by the following expression:

$$C = 10\log\left(\frac{P_1}{P_3}\right) \tag{2.29}$$

with $P_1$ the incident power at port 1 and $P_3$ the output power at port 3.

In the ideal case, the power that exits via port 4 is zero, and therefore this is called the "isolated port." In practical realizations, the isolation is not perfect, and is characterized by the following expression:

$$I = 10\log\left(\frac{P_1}{P_4}\right) \tag{2.30}$$

with $P_1$ the incident power at port 1 and $P_4$ the output power at port 4.

In this way, the incident signal applied to the DUT can be measured.

A directional coupler also allows one to measure the scattered wave. In this case, port 2 acts as the input port of the directional coupler, because the wave scattered by the DUT enters the directional coupler at port 2. Now port 4 is the coupled port and port 3 is the isolated port. The directional coupler's ability to separate forward and backward waves is expressed by the directivity:

$$D = 10\log\left(\frac{P_3}{P_4}\right) \tag{2.31}$$

The above formulas are related by the following expression:

$$I = D + C \tag{2.32}$$

The non-perfect isolation can be corrected for during the calibration, which will be discussed in Section 2.6.2.

#### 2.6.1.3 The receiver

The key component of the architecture is the receiver that measures the incident and scattered traveling voltage waves at each port. In the early years of network analyzers, only the magnitude could be measured, and therefore the instrument was called a "scalar" network analyzer. Nowadays, all network analyzers measure both amplitude and phase, which explains the name "vector" network analyzer.

The receiver is based on a tuned configuration, as in the case of spectrum analyzers (Section 2.3). The topology is usually the superheterodyne architecture, as can be seen in Fig. 2.49. The configuration has three blocks: a mixer, a filter, and an analog-to-digital converter. The input signal is down-converted to a lower intermediate frequency (IF) by mixing the input signal with the local oscillator (LO). The LO is locked to either the RF or the IF signal so that the receivers in the network analyzer are always tuned to the RF signal present at the input. The IF signal is bandpass filtered, mainly to remove unwanted signals created in the mixer, thereby improving the sensitivity and dynamic range. Note that, in the case of the VNA, a two-receiver configuration is needed so that one can measure the amplitude and phase of a scattered wave relative to an incident wave, or, in other words, obtain the $S$-parameters. After filtering, the amplitude can be

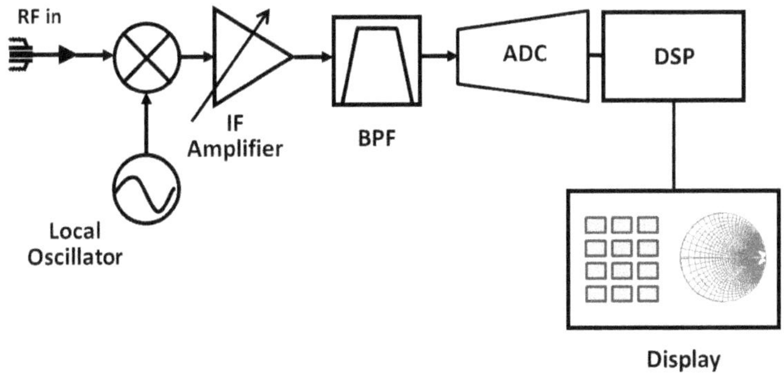


**Figure 2.49** The receiver of a vector network analyzer.

measured using an envelope detector, and the phase may be measured with a quadrature detector. However, modern network analyzers use an analog-to-digital converter (ADC) and digital signal processing (DSP) to extract magnitude and phase information from the IF signal.

Finally the measured $S$-parameters are displayed in a format that can be interpreted, such as a Smith-chart plot.

For the overall configuration to work properly, the instrument should be carefully calibrated prior to performing any measurement, which is the subject of the next section.

## 2.6.2 Calibration

The quantities in which we are interested are the waves at the DUT's ports, indicated by superscript D in Fig. 2.47. The instrument is, however, acquiring the raw quantities, indicated by superscript R. The aim of the process of calibration is to remove the systematic errors between the connector ports, or probe tips in the case of on-wafer measurements, and the actual data acquisition. This includes effects such as losses and phase shifts in cables, imperfect directional couplers, mismatches, etc. In general terms, the relationship between the four "DUT" waves and the four "raw" waves can be expressed by a $4 \times 4$ matrix:

$$\begin{bmatrix} a_{1i}^{\mathrm{D}} \\ b_{1i}^{\mathrm{D}} \\ a_{2i}^{\mathrm{D}} \\ b_{2i}^{\mathrm{D}} \end{bmatrix} = \begin{bmatrix} \alpha'_{1i} & \beta'_{1i} & 0 & 0 \\ \gamma'_{1i} & \delta'_{1i} & 0 & 0 \\ 0 & 0 & \alpha'_{2i} & \beta'_{2i} \\ 0 & 0 & \gamma'_{2i} & \delta'_{2i} \end{bmatrix} \begin{bmatrix} a_{1i}^{\mathrm{R}} \\ b_{1i}^{\mathrm{R}} \\ a_{2i}^{\mathrm{R}} \\ b_{2i}^{\mathrm{R}} \end{bmatrix} \tag{2.33}$$

with $i$ indicating the frequency index.

The aim of the calibration process is to determine the unknown matrix coefficients. It can be observed that 8 of the 16 coefficients in Eq. (2.33) are set equal to zero. This corresponds to the assumption that there is no coupling between ports 1 and 2. In other words, for example, $a_{1i}^{\mathrm{D}}$ can be determined from the raw quantities measured at port 1, namely $a_{1i}^{\mathrm{R}}$ and $b_{1i}^{\mathrm{R}}$, and has no contribution from the raw quantities measured at port 2. This assumption is usually valid for connectorized measurements in the lower microwave range, and hence covers the wireless standards bands. In the case of on-wafer measurements, this assumption is less applicable due to the short distance between the probe tips, which increases the coupling. Since this book primarily focuses on connectorized measurements, we assume that the no-coupling assumption is valid in the following.

In the case of $S$-parameter measurements, we are interested only in ratios of waves, as explained above in Section 2.6.1. Therefore we can normalize the matrix coefficients with respect to, for example, the first coefficient $\alpha'_{1i}$. As a result, only seven frequency-dependent coefficients need to be determined by the calibration procedure. The complex number $\alpha'_{1i}$ does not have to be determined in the $S$-parameter-calibration procedure, since this number will be eliminated on taking the ratios corresponding to the $S$-parameters. On the other hand, determining $\alpha'_{1i}$ is essential in the NVNA calibration, as will be explained in Section 2.7. Thus

$$\begin{bmatrix} a_{1i}^{\mathrm{D}} \\ b_{1i}^{\mathrm{D}} \\ a_{2i}^{\mathrm{D}} \\ b_{2i}^{\mathrm{D}} \end{bmatrix} = \alpha'_{1i} \begin{bmatrix} 1 & \beta_{1i} & 0 & 0 \\ \gamma_{1i} & \delta_{1i} & 0 & 0 \\ 0 & 0 & \alpha_{2i} & \beta_{2i} \\ 0 & 0 & \gamma_{2i} & \delta_{2i} \end{bmatrix} \begin{bmatrix} a_{1i}^{\mathrm{R}} \\ b_{1i}^{\mathrm{R}} \\ a_{2i}^{\mathrm{R}} \\ b_{2i}^{\mathrm{R}} \end{bmatrix} \tag{2.34}$$

There exist many calibration procedures for $S$-parameter measurements. The choice is essentially determined by the validity of the underlying assumptions. In this book, we focus on the so-called SOLT calibration, which is the most commonly adopted approach for connectorized measurements in the frequency range for wireless applications, as well as on TRL calibration, since this is useful when one wants to develop one's own calibration kit.

#### 2.6.2.1 SOLT calibration

The letters of SOLT refer to the standards being used, namely short, open, load, and through. The load has a resistance of 50 Ω. The standards are supplied by manufacturers, grouped in a calibration box (Fig. 2.50). Lately, also the so-called e-cal has become possible. This involves a small box that the user has to connect only once, and then the various standards are applied sequentially by means of internal switches. Here we will explain the manual process.

The sequence in which the standards are connected is not important. Usually, one connects first the three one-port standards (open, short, and load) sequentially to each of the ports, and then the two ports are interconnected with the "through" standard. The standards are not perfect, and therefore the manufacturer supplies their actual reflection coefficient Γ. For example, the "open" standard has a parasitic capacitance value, and the load has a residual inductance value due to the short transmission line connecting the connector to the internal 50-Ω resistor. This reflection coefficient Γ is frequency-dependent.

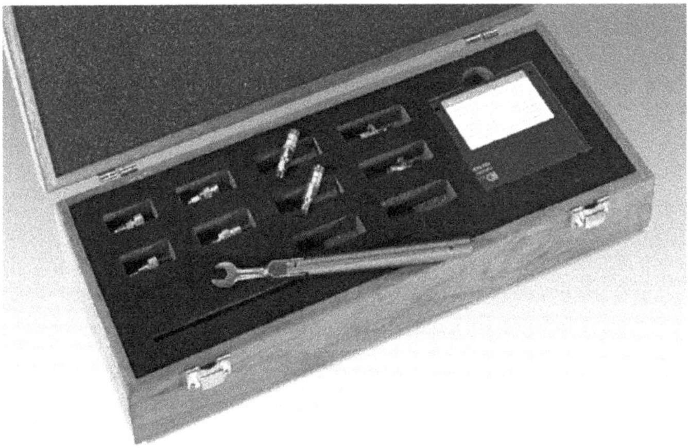

**Figure 2.50** A calibration box. © Agilent.

The standards are applied at the connector ports, and, since this corresponds to the "DUT" reference plane, we can write the following expressions:

$$\Gamma_i^{\mathrm{O}} a_{1i}^{\mathrm{D1}} = b_{1i}^{\mathrm{D1}} \tag{2.35}$$

$$\Gamma_i^{\mathrm{S}} a_{1i}^{\mathrm{D2}} = b_{1i}^{\mathrm{D2}} \tag{2.36}$$

$$\Gamma_i^{\mathrm{L}} a_{1i}^{\mathrm{D3}} = b_{1i}^{\mathrm{D3}} \tag{2.37}$$

$$\Gamma_i^{\mathrm{O}} a_{2i}^{\mathrm{D4}} = b_{2i}^{\mathrm{D4}} \tag{2.38}$$

$$\Gamma_i^{\mathrm{S}} a_{2i}^{\mathrm{D5}} = b_{2i}^{\mathrm{D5}} \tag{2.39}$$

$$\Gamma_i^{\mathrm{L}} a_{2i}^{\mathrm{D6}} = b_{2i}^{\mathrm{D6}} \tag{2.40}$$

The superscript numbers 1 to 6 number the measurements, with O for open, S for short, and L for load.

Next, we can express the DUT quantities in terms of the measured quantities by applying Eq. (2.34). Note that the factor $\alpha'_{1i}$ has been omitted from the expressions, since it will vanish anyhow on taking the ratios for the $S$-parameters. Thus

$$\Gamma_i^{\mathrm{O}} \left( a_{1i}^{\mathrm{R1}} + \beta_{1i} b_{1i}^{\mathrm{R1}} \right) = \gamma_{1i} a_{1i}^{\mathrm{R1}} + \delta_{1i} b_{1i}^{\mathrm{R1}} \tag{2.41}$$

$$\Gamma_i^{\mathrm{S}} \left( a_{1i}^{\mathrm{R2}} + \beta_{1i} b_{1i}^{\mathrm{R2}} \right) = \gamma_{1i} a_{1i}^{\mathrm{R2}} + \delta_{1i} b_{1i}^{\mathrm{R2}} \tag{2.42}$$

$$\Gamma_i^{\mathrm{L}} \left( a_{1i}^{\mathrm{R3}} + \beta_{1i} b_{1i}^{\mathrm{R3}} \right) = \gamma_{1i} a_{1i}^{\mathrm{R3}} + \delta_{1i} b_{1i}^{\mathrm{R3}} \tag{2.43}$$

$$\Gamma_i^{\mathrm{O}} \left( a_{2i}^{\mathrm{R4}} + \beta''_{2i} b_{2i}^{\mathrm{R4}} \right) = \gamma''_{2i} a_{2i}^{\mathrm{R4}} + \delta''_{2i} b_{2i}^{\mathrm{R4}} \tag{2.44}$$

$$\Gamma_i^{\mathrm{S}} \left( a_{2i}^{\mathrm{R5}} + \beta''_{2i} b_{2i}^{\mathrm{R5}} \right) = \gamma''_{2i} a_{2i}^{\mathrm{R5}} + \delta''_{2i} b_{2i}^{\mathrm{R5}} \tag{2.45}$$

$$\Gamma_i^{\mathrm{L}} \left( a_{2i}^{\mathrm{R6}} + \beta''_{2i} b_{2i}^{\mathrm{R6}} \right) = \gamma''_{2i} a_{2i}^{\mathrm{R6}} + \delta''_{2i} b_{2i}^{\mathrm{R6}} \tag{2.46}$$

with

$$\beta''_{2i} = \frac{\beta_{2i}}{\alpha_{2i}} \tag{2.47}$$

$$\gamma''_{2i} = \frac{\gamma_{2i}}{\alpha_{2i}} \tag{2.48}$$

$$\delta''_{2i} = \frac{\delta_{2i}}{\alpha_{2i}} \tag{2.49}$$

The above expressions already provide six equations to determine the unknown matrix coefficients. In order to determine all seven unknowns, we also need a seventh measurement, which is the measurement of the "through" standard:

$$a_{1i}^{\mathrm{D7}} = b_{2i}^{\mathrm{D7}} \tag{2.50}$$

In this expression, it is assumed that the "through" connection is perfect, meaning that the scattered wave at port 2, $b_{2i}^{\mathrm{D7}}$, is equal to the incident wave $a_{1i}^{\mathrm{D7}}$ at port 1. In reality, the "through" standard will have a delay, or phase shift, and also a small loss. This imperfection is incorporated into the built-in calibration algorithm in the VNA.

As with the other standards' measurements, we can substitute for the "DUT" quantities the actually measured quantities:

$$a_{1i}^{\mathrm{R7}} + \beta_{1i} b_{1i}^{\mathrm{R7}} = \alpha_{2i}\left(\gamma''_{2i} a_{2i}^{\mathrm{R7}} + \delta''_{2i} b_{2i}^{\mathrm{R7}}\right) \tag{2.51}$$

Now we have seven equations, so all of the unknown calibration coefficients can be determined. This concludes the SOLT calibration procedure.

#### 2.6.2.2 TRL calibration

The TRL calibration method was introduced by Engen [11]. The letters of TRL refer again to the standards being used, namely through, reflect, and line. It is an alternative technique to SOLT, especially when the instrument manufacturer's connectorized calkit or, in the case of on-wafer measurements, calibration substrate is not applicable for the device to be measured. The typical example is measuring devices in test fixtures. SOLT requires an accurately trimmed load. This is a challenge when one has to develop one's own calibration kit. TRL is less stringent fabricationwise. It does not require a load, the reflect doesn't have to be accurately known, and the characteristic impedance $Z_0$ of the through and line should be equal, but the value does not have to be known. More precisely, the TRL standards have the following requirements.

- Through. $Z_0$ should be the same as $Z_0$ of the line standard. The attenuation doesn't have to be known. If it is a non-zero through, the electrical length should be well known and specified if the through is going to be used to set the reference plane.
- Reflect. the magnitude of the reflection coefficient is optimally 1.0, but it is not necessary to know the exact value. The phase of the reflection coefficient must be known to within 90° accuracy. However, if the reflect is used to set the reference plane, its phase response must be well known and specified. Also, the reflect must be identical on both ports. Typically one realizes either a short circuit or an open circuit.

**Figure 2.51** A TRL calibration kit for in-fixture measurements.

- Line. the characteristic impedance $Z_0$ of the line establishes the impedance of the measurement (i.e., $S_{11} = S_{22} = 0$). The insertion phase of the line must be different from that of the through. The electrical length difference between the through and the line must be larger than 20° and smaller than 160°. The attenuation doesn't need to be known. The electrical length should be known.

Owing to the requirement that the electrical length difference between the through and the line must be between 20° and 160°, the bandwidth of the calibration might not be sufficient for the targeted measurements. In such cases, a set of multiple lines should be used.

Modern VNAs support TRL calibrations (Fig. 2.51) and have built-in procedures that ask the operator to connect the various standards, and then calculate the calibration coefficients.

### 2.6.3 VNA measurement uncertainty

Measurement uncertainty in the case of $S$-parameter measurements is more complex [12] than for the instruments discussed in previous sections. To acquire a basic understanding of measurement uncertainty, the reader is referred to the comprehensive online "Guide to the expression of uncertainty in measurement," known as the GUM document, and its appendices [13].

## 2.7 Nonlinear vector network analyzers

Whereas a vector network analyzer returns only ratios of waves, absolute measurements can be obtained with a nonlinear vector network analyzer (NVNA). In this book,

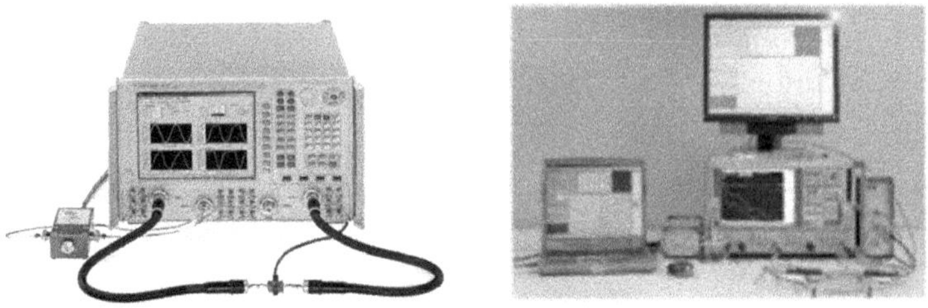

**Figure 2.52** Two mixer-based NVNAs. Images © external vendors.

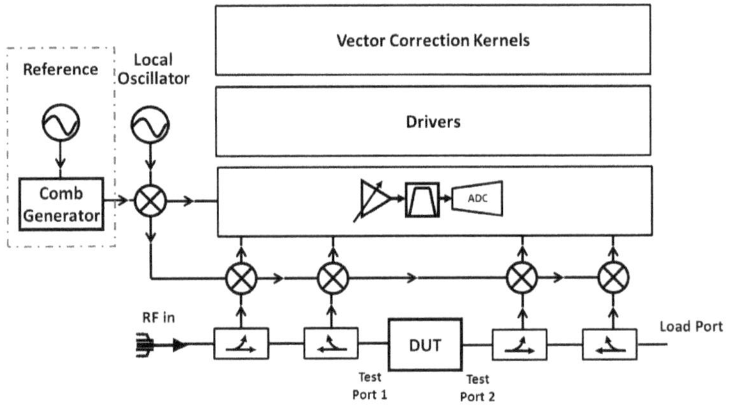


**Figure 2.53** The mixer-based NVNA architecture [15] © IEEE.

we focus on the architecture that is most widespread nowadays, namely the so-called mixer-based approach (Fig. 2.52). This approach was preceded by several alternative architectures making use of harmonic sampling (i.e., the microwave transition analyzer (MTA) and derived architectures such as the large-signal network analyzer (LSNA)) and equivalent time sampling (i.e., oscilloscopes). For a historical overview, see [14].

## 2.7.1 Architecture

The mixer-based NVNA architecture is illustrated in Fig. 2.53. It is based on a four-port VNA [15]. In fact, a five-channel VNA would be sufficient, but the typical commercial VNA implementations are either two-port or four-port. Four channels, or two ports, are needed in order to acquire the incident and traveling voltage waves at the two-port DUT's reference plane. The fifth channel is used to enable the harmonic phase measurements. In a normal VNA configuration, the local oscillator frequency is swept across the measurement frequency range. Since the phase of the local oscillator varies from measurement to measurement, the phase relationships of the incident and scattered waves at the harmonic frequencies cannot be determined. The use of the fifth channel resolves this problem, since it enables one to characterize the phase of the local oscillator at each frequency, and thus it becomes possible to characterize also the harmonic

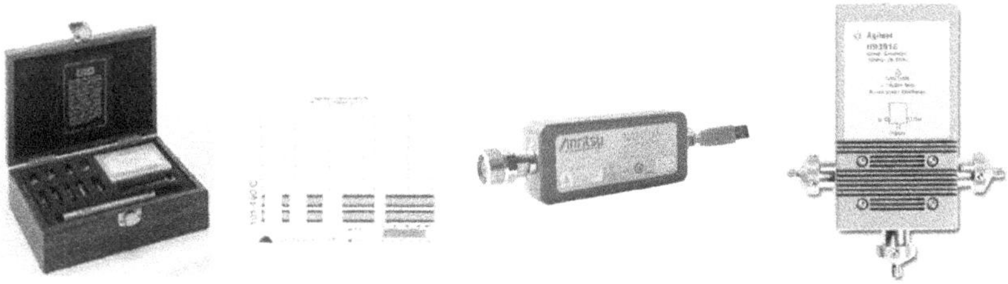

**Figure 2.54** Calibration elements for NVNA measurements: from left to right, calibration box, on-wafer calibration substrate, power sensor, and harmonic phase reference. Images © external vendors.

phase relationships of the incident and scattered waves at the DUT's reference plane. The phase of the local oscillator is tracked by means of a so-called harmonic phase reference device (Fig. 2.54).

Even though the internal architecture of this harmonic phase reference device has changed over the years, especially to cover wider bandwidths, the principle is that it behaves as a comb generator. This means that it is a strongly nonlinear device, such that, when excited by a reference signal, e.g., at 10 MHz, the output signal has a large number of harmonics with high amplitudes. The reference frequency at which the device is excited is chosen to be the highest common factor of all frequencies in the signals present at the measurement ports, or a subharmonic of that frequency. One harmonic of the reference frequency will then fall at each frequency of interest. Since it is assumed that the absolute phase relationship of each comb tone remains static for the duration of a measurement, the respective phase of the local oscillator at each harmonic frequency can be determined, and thus so can the relative phase relationships of the measured DUT quantities.

## 2.7.2 Calibration

The aim of the calibration process is similar to that with $S$-parameter measurements in the sense that we want to determine the "DUT" quantities from the measured "raw" quantities. We reconsider the matrix relationship as introduced in Eq. (2.34). The complex factor $\alpha'_{1i}$ is now represented by its magnitude $K_i$ and phase $\Phi_i$:

$$\begin{bmatrix} a_{1i}^{\mathrm{D}} \\ b_{1i}^{\mathrm{D}} \\ a_{2i}^{\mathrm{D}} \\ b_{2i}^{\mathrm{D}} \end{bmatrix} = K_i e^{j\Phi_i} \begin{bmatrix} 1 & \beta_{1i} & 0 & 0 \\ \gamma_{1i} & \delta_{1i} & 0 & 0 \\ 0 & 0 & \alpha_{2i} & \beta_{2i} \\ 0 & 0 & \gamma_{2i} & \delta_{2i} \end{bmatrix} \begin{bmatrix} a_{1i}^{\mathrm{R}} \\ b_{1i}^{\mathrm{R}} \\ a_{2i}^{\mathrm{R}} \\ b_{2i}^{\mathrm{R}} \end{bmatrix} \tag{2.52}$$

In the case of NVNA measurements, the aim is to determine the absolute value of the waves, and therefore $K_i$ and $\Phi_i$ must now be determined as well. This also implies that the coefficients in the matrix are still being determined using an $S$-parameter calibration technique. In summary, the calibration consists of three steps:

(1) linear calibration
(2) power calibration
(3) harmonic phase calibration

For the linear calibration, the $S$-parameter calibration technique of preference can be adopted. As mentioned in Section 2.6.2, the calibration technique of choice depends on the target frequency range and whether the measurements are connectorized or on-wafer. To determine the frequency-dependent magnitude $K_i$, a frequency-dependent power calibration using a power meter, as described in Section 2.2, should be carried out. In order to determine the unknown phase $\Phi_i$, a phase reference is used [16]. This implies that in total two phase references are required in the case of the mixer-based architecture: one phase reference to track continuously the phase of the frequency-sweeping local oscillator and a second phase reference for the calibration. To perform the NVNA phase calibration, the phase reference is connected to the instrument. It is excited by the lowest frequency of interest, typically 10 MHz. The rich harmonic content of the phase reference enables one to determine the unknown phase relationships $\Phi_i$ for harmonics $i$ of the excitation signal up to the millimeter-wave range. The phase reference on its own can be calibrated by connecting it to a sampling oscilloscope. The wide harmonic frequency spectrum corresponds to a sharp pulse in the time domain. Since the measurement of a sampling oscilloscope is traceable to physical quantities [17], the measurement of the phase reference is traceable. Such measurement of the phase reference is usually executed by the manufacturer, who has access to a traceable sampling oscilloscope.

Figure 2.54 illustrates the set of calibration elements for an NVNA calibration. The calibration box or on-wafer reference substrate is used for connectorized or on-wafer linear calibration, respectively. The power sensor and phase reference are connectorized and do not yet have an on-wafer equivalent. For on-wafer NVNA measurements, an additional, manufacturer-dependent calibration step is required in order to refer the power and harmonic phase calibrations to the on-wafer reference plane.

The uncertainty description of NVNA measurements is still an on-going research topic. As with the calibration, it can be subdivided into three parts, namely the linear measurements' uncertainty, power measurement uncertainty, and absolute phase measurement uncertainty. The linear measurements' uncertainty can be referred to the uncertainty in $S$-parameter measurements (see Section 2.6.2), whereas the power measurement uncertainty can be linked to the uncertainty in power-meter measurements (see Section 2.2.5). The uncertainty of the phase standard is still being investigated [18].

## 2.8 Oscilloscopes

The figures of merit presented in chapter 1 are mostly frequency-domain-based ones. An oscilloscope is a time-domain-based instrument. Even though the FFT can be used to obtain the corresponding frequency spectrum, most wireless engineers prefer to work with instrumentation such as SAs and VSAs. One of the reason for this is the lower dynamic

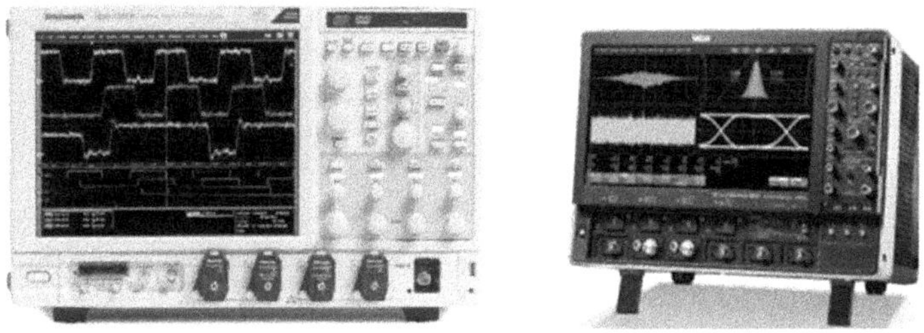

**Figure 2.55** The oscilloscopes. Images © external vendors.

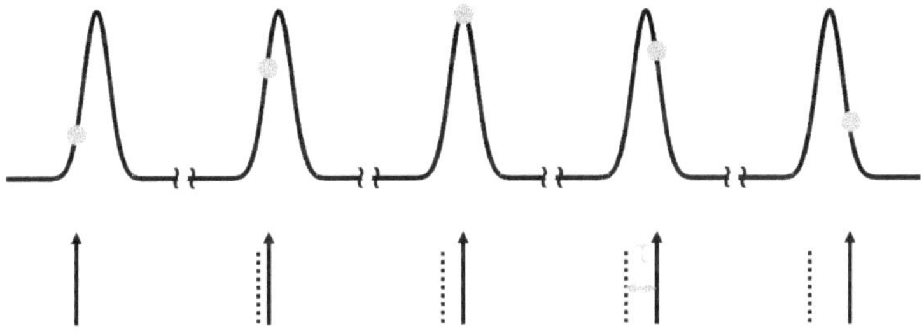

**Figure 2.56** The operating principle of an equivalent-time sampling oscilloscope.

range, and also the frequent need for repeat calibrations due to time-base distortion, jitter, and drift. Nevertheless, it can be useful to have an oscilloscope in the lab, for example for quick de-bugging purposes. Some examples of high-frequency oscilloscopes are depicted in Fig. 2.55.

A distinction should be made between real-time oscilloscopes and equivalent-time sampling oscilloscopes. Real-time oscilloscopes sample a waveform at a very high rate and store the measurements in a circular memory buffer. By setting trigger events (see later), the user can determine which part of the waveform to view. The fundamental advantage of a real-time oscilloscope is that it enables one to characterize one-time transient effects. On the other hand, since it requires a very fast ADC, the bandwidth of a real-time oscilloscope is substantially less than that of an equivalent-time sampling oscilloscope.

The operating principle of an equivalent-time sampling oscilloscope, often called a sampling oscilloscope, is represented in Fig. 2.56. One assumes that the signal is periodic, implying that single-shot events cannot be measured with this instrument. The moment at which a sample is taken is varied from cycle to cycle by means of the varying delay $\tau$, such that in the end the full waveform has been characterized. The typical bandwidth is on the order of 70 GHz.

An important parameter in oscilloscope measurements is triggering. The trigger event defines the point in time at which a repeating window of waveform information is stabilized for viewing. The trigger also allows one to capture single-shot waveforms in the case of real-time oscilloscopes. Edge triggering is the basic and most common type. For edge triggering, the user selects the slope (positive or negative, or both) and level, and

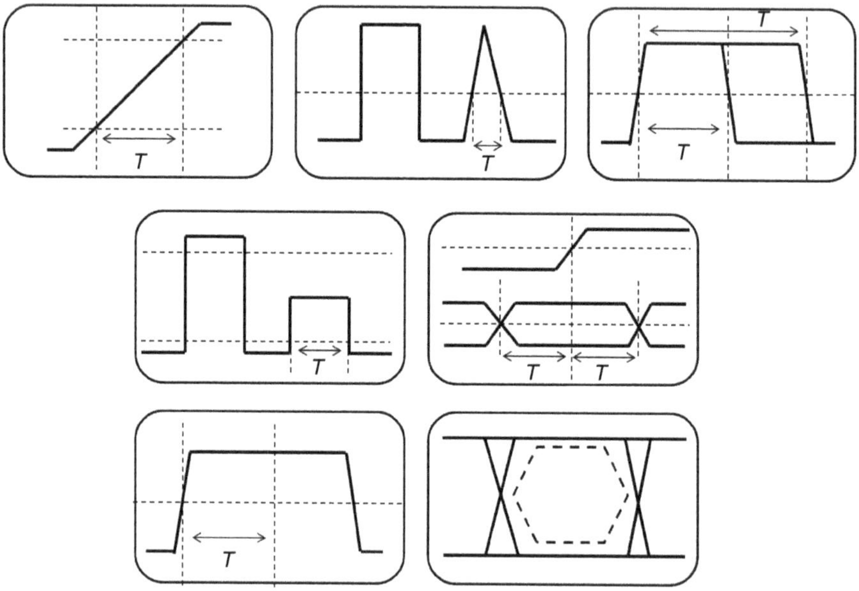


**Figure 2.57** Varieties of oscilloscope triggering: from left to right, slew-rate triggering, glitch triggering, pulse-width triggering, runt-pulse triggering, setup-and-hold triggering, time-out triggering, and communication triggering.

the oscilloscope triggers when the signal meets these conditions. This is also known as threshold crossing. The types of trigger control are diverse and manufacturer-dependent. One can trigger on signals on the basis of amplitude on time, and even by logic state or pattern. Some examples are listed in the following overview, and graphically illustrated in Fig. 2.57.

- Slew-rate triggering: trigger on fast or slow edges. This approach is more advanced than edge triggering, insofar as the latter does not consider the slope itself, but only the threshold value.
- Glitch triggering: trigger on pulses when they are shorter or longer than a user-defined time limit.
- Pulse-width triggering: accept (or reject) only those triggers defined by pulse widths that are between two defined time limits. This is useful for observing inter-symbol interference (ISI).
- Runt-pulse triggering: accept only those triggers defined by pulses that enter and exit between two defined amplitude thresholds.
- Setup-and-hold triggering: trap a single violation of setup-and-hold time that would almost certainly be missed by using other trigger modes. This trigger mode makes it easy to capture specific details of signal quality and timing when a synchronous data signal fails to meet setup-and-hold specifications.
- Time-out triggering: trigger on an event without waiting for the trigger pulse to end, by triggering on the basis of a specified time lapse.

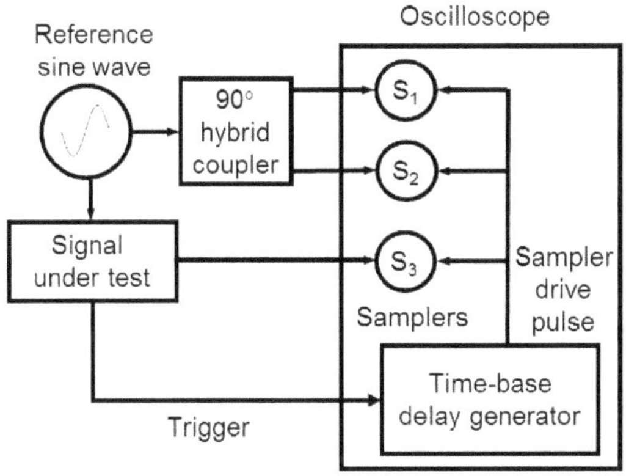


**Figure 2.58** The setup for time-base error correction [19] © IEEE.

- Logic triggering: trigger on any logical combination of available input channels. This approach is especially useful in verifying the operation of digital logic.
- Communication triggering: these trigger modes address the need to acquire a wide variety of alternate-mark inversion (AMI), code-mark inversion (CMI), and non-return-to-zero (NRZ) communication signals.

Finally, to correct for time-base errors in oscilloscopes, the setup shown in Fig. 2.58 can be applied. The idea is that a sine wave is measured simultaneously with the signal to be characterized. By phase-shifting the sine wave by 0° and 90°, the necessary information to enable error correction can be obtained. We refer the reader to [19] for the mathematical details. It is important to realize that this method requires the oscilloscope to have at least three channels. Also, the source generating the reference sine wave and the signal to be characterized should be time-synchronized.

## 2.9 Logic analyzers

Logic analyzers are similar to digital oscilloscopes, but, rather than capturing the analog signal itself, they capture logic states in a predetermined input. By logic states we mean digital signals, which can be translated as follows: if a signal is above a certain voltage threshold $V_{th}$ then a “1” is measured, whereas, if it is below the threshold, a “0” is measured. This is illustrated in Fig. 2.59.

The reader might wonder why we are talking about logic analyzers in this book, but, if we remember that emerging wireless systems are rapidly tending towards being all digital, either at the receiving stage or at the transmitter stage, then the radio path will be a bus (with $n$ bits) that contains a digital word corresponding to the signal waveform. With this concept of a digital signal in mind, a logic analyzer is nothing more than a very large number of time scopes, one for each input.

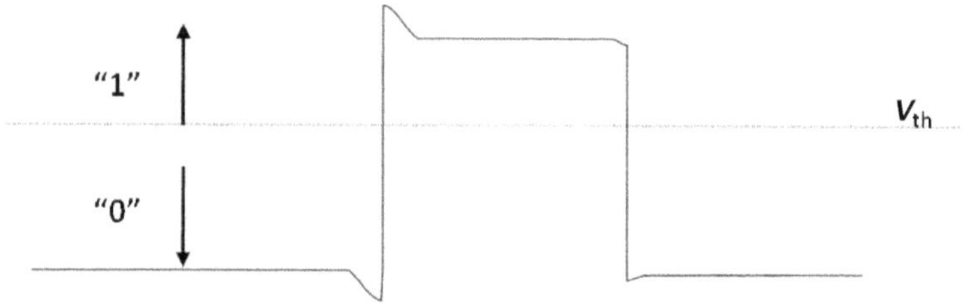


**Figure 2.59** Logic states in a logic analyzer.

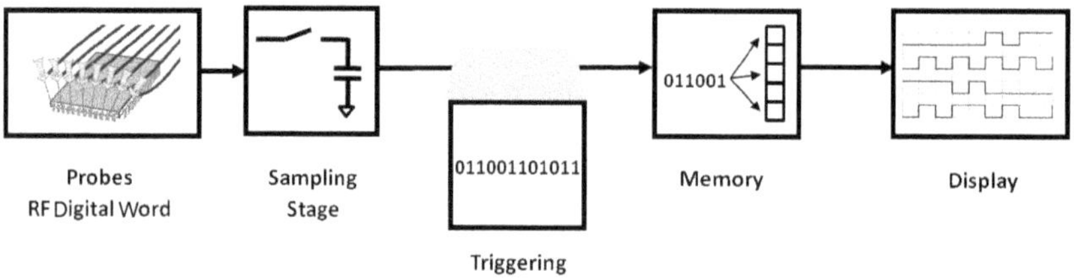


**Figure 2.60** Logic-analyzer architecture.

**Figure 2.61** Logic-analyzer probes. © Tektronix.

In contrast to the case of an oscilloscope, in a logic analyzer the number of inputs corresponds to the number of bits we can gather from our DUT, which in this case we call the system under test (SUT). The typical number of inputs spans from near 34 to 136 lines in commercial logic analyzers.

Figure 2.60 presents a typical logic-analyzer architecture. As can be seen, the architecture contains a probing stage, a sampling stage, a huge real-time memory, a triggering mechanism, and a display for gathering all the information. These blocks will be explained individually.

## 2.9.1 Logic-analyzer probes

Regarding the probing stage, it is important to consider that in this case the objective is to acquire several signals that can have very high speed, so a correct probing mechanism is fundamental. Figure 2.61 presents some of the typical probes used in logic-analyzer measurements.

We can identify three main types of probes. The first type consists of the so-called general-purpose probes. Such probes are based on "flying lead sets," which are so called

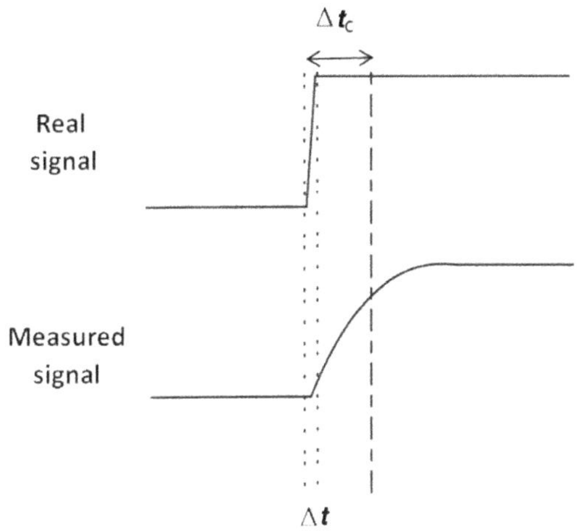


**Figure 2.62** Probe capacitive problems.

because they are actually leads that do not contact each other, so they can actually fly over the circuit to be measured, and are intended for picking up point-by-point samples, just like a conventional oscilloscope probe. These probes are used mainly for troubleshooting purposes.

Another probe type consists of high-density, multi-channel probes. In this case the probe is packed into a specific connector that can be hooked to a connector pair, which is normally embedded into the circuit to be measured.

Sometimes it is also recommended that probes of a third type, namely connector-less high-density probes, be used.

It is important to note that, despite the fact that logic analyzers have to measure logic states, the actual signal is a voltage waveform that changes with time, and, in the case of wireless signals, that variation over time can be extremely fast. Thus any mismatch at each probe point can actually degrade the signal itself. This is especially problematic due to the capacitances of the probes. Figure 2.62 presents this capacitive problem, where the logic state, which should be a pure square wave, is being delayed due to the capacitive phenomena. Thus an error can occur when identifying the correct logic state, and then be attributed to the SUT, whereas the error actually arises from a badly chosen probe.

It is recommended that, before starting a logic analysis, the user identify the correct probe in terms of maximum operation frequency, in order not to degrade the measurement and corrupt it with noise.

Despite this, for high-frequency signals it is also fundamental, and obligatory, that each probe should have a ground line, and most of the time the same cable should combine a line and a ground path simultaneously in order to reduce significantly the noise. If this procedure is not implemented then the noise can corrupt some of the transmitted signals.

**Table 2.6** Logic-analyzer triggers

| Event | Description |
|---|---|
| Words | One specific association of bits defining a word |
| Signal | In this case any external signal that acts as a reset, for instance |
| Counter | Some internal counting mechanism pre-programmed by the user |
| Ranges | Some events that occur between a low and a high value |
| Glitches | Pulses that can appear between acquisitions |
| Analog | Use an oscilloscope to trigger the digital part on the basis of an analog event |
| Timer | Elapsed time between two events |

### 2.9.2 Sampling the logic data

The second block to account for in the logic analyzer is the sampling stage. Sometimes manufacturers call this process the clock mode. We can consider two different types of sampling, namely the timing acquisition (sometimes called asynchronous acquisition) and the state acquisition (sometimes called synchronous acquisition).

The timing mode is just like a sampling device. The logic analyzer measures a new value each sampling interval, imposed by an internal clock, and thus any event that happens in between is ignored. This will fill up the memory with logic patterns that vary with time. The smallest sampling time corresponds to the highest bit rate we can measure.

The other sampling mechanism is the state acquisition. It is also called synchronous acquisition since the logic analyzer will behave as a synchronous device that captures data when a certain state is activated. The state can be the system clock, a control signal on the bus, or any other signal that causes the SUT to change states.

In the sampling mechanism we can also define when the sampling is done, i.e., during the falling or during the rising edge. In the state mode the signal should be stable prior to the measurement, in order to guarantee a good level of measurement confidence.

It is also useful that the probes are able to acquire simultaneously the state and the timing, since that will facilitate a huge amount of problem detection and will significantly improve the de-bugging time.

### 2.9.3 Triggering

Triggering in a logic analyzer has a broader meaning than in an oscilloscope, since we can define a multitude of triggering forms. To understand triggering in logic analyzers, we must define an "event." An "event" is a signal pattern that can initiate a trigger; for instance, an "event" can be a glitch in the bus, a specific word, or any other form of signal pattern. Table 2.6 presents some forms of triggering.

It is important to mention that the better the time resolution, the better the triggering that can be achieved.

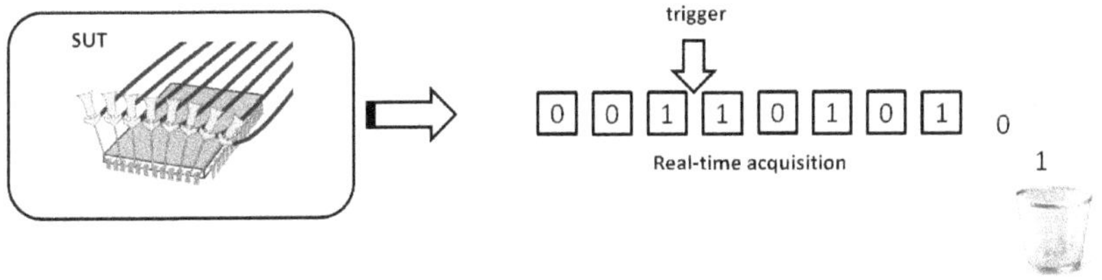


**Figure 2.63** Memory filling up.

## 2.9.4 Real-time memory

After the signals have been sampled and acquired, they will fill up a real-time memory. Actually it is this memory which will contain all the measured data and will allow the user to understand the functioning of the SUT, and thus its subsequent characterization, for instance in terms of the spectrum, BER, EVM, etc.

The memory can start to be filled up at a specified trigger, and this process can be stopped at another trigger. In this way the user can analyze the overall acquisition very efficiently. The user can afterwards decide what to do with the saved data.

Figure 2.63 shows how the memory becomes filled up, as well as some triggering mechanisms.

## 2.9.5 Analyzing the acquired signal

The signal that fills up the memory can then be analyzed either at a digital level, so we can see what the bus contains, or as a representation of the analog level. In the latter case the digital words are converted back to a representation of the analog signal by using a simple mathematical expression like

$$x_{\text{analog}_{\text{rep}}}(t) = D_{0_{\text{LSB}}}2^0 + D_1 2^1 + D_2 2^2 + \cdots + D_{N_{\text{MSB}}}2^N \tag{2.53}$$

where LSB is the least significant bit, MSB the most significant bit, $N$ the number of bits, and $D_0$, ..., $D_N$ the bit stream. Using this equivalence, the user can plot the evolution of the signal over time, can calculate the frequency behavior using an FFT, or can plot the constellation diagram of a modulated complex envelope signal, and can implement all the types of measurement that are adopted with analog signals, as illustrated in Fig. 2.64.

In systems of this type the accuracy and uncertainty aspects are similar to those in the previously presented approach involving digital oscilloscopes.

# 2.10 Noise-figure measurement

In the past, when no digital scopes were available, one encountered noise-figure (NF) meters as standalone instruments that included noise sources and measurement scopes within the same box. Nowadays NF measurement is not done using an instrument by itself, but is dealt with by a firmware update containing software that is able to calculate

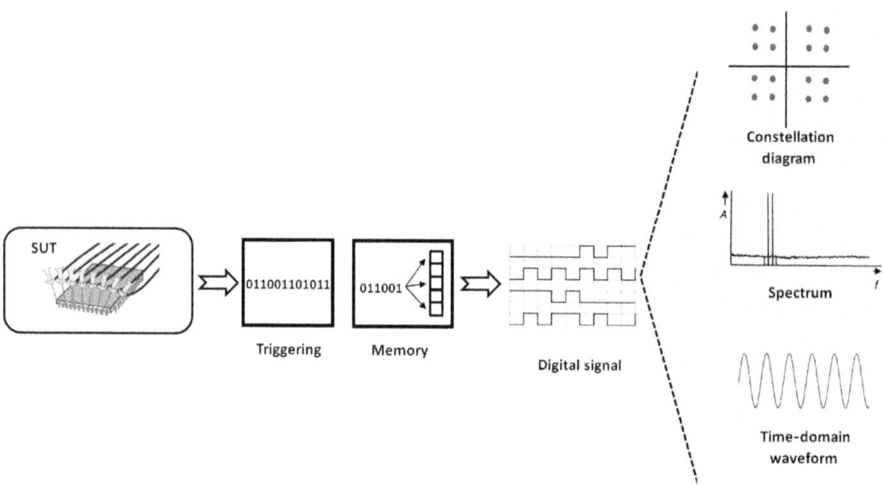


**Figure 2.64** Displaying the information using a digital word.

the NF values from different measurements of power, normally with and without a noise source, or, in certain implementations, with the noise source connected (hot) and disconnected (cold).

The NF can thus be measured using a spectrum analyzer (or, in certain contexts, a power meter) or, in new instruments, using a vector network analyzer. With a spectrum analyzer, the main idea is to use a noise source, usually implemented using a diode, and to measure the power with the noise source on and off. With a network analyzer the main idea is to measure a single sinusoid by taking several averages. The average value is then compared with the non-averaged value, and the difference is the added noise as was explained in Section 1.3.2. The two implementations are now discussed in more detail.

## 2.10.1 Noise-figure measurement using a noise source

The traditional way to measure the noise factor (noise figure) is by using a noise source, and measuring the power at the output of the DUT with the noise source on and off. This is illustrated in Fig. 2.65.

This method involves first a calibration of the system, by measuring the noise source directly connected to the spectrum analyzer, as will be seen in Section 4.2.2. This first step will impose the calibration of the system, and will measure the noise power of the noise source when it is connected, $P_{\mathrm{n}_{\mathrm{generatorON}}}$, and disconnected, $P_{\mathrm{n}_{\mathrm{generatorOFF}}}$.

In the next step the DUT is inserted between the noise generator and the SA, Fig. 2.65. Again the noise source will be connected and disconnected, and the power appearing in the SA will be measured. The output noise of a DUT, when no noise source is applied at the input, is given by

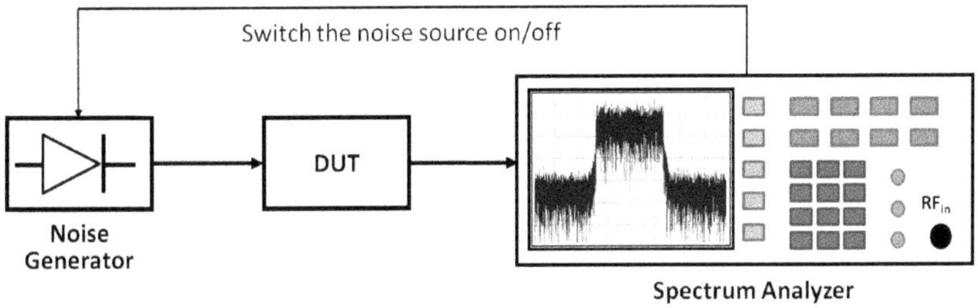


**Figure 2.65** Noise-figure measurement using a noise source.

$$P_{\text{nDUTOFF}} = k_{\text{B}} T B F_{\text{overall}} G_{\text{DUT}} \tag{2.54}$$

as was presented in Section 1.3.2. $F_{\text{overall}}$ corresponds to the overall noise factor of the DUT and the spectrum analyzer in this case.

If we now add a noise source at the input then the output power will be

$$P_{\text{nDUTON}} = G_{\text{DUT}} k_{\text{B}} T B F_{\text{overall}} + G_{\text{DUT}} k_{\text{B}} T B F_{\text{noisesource}} \tag{2.55}$$

In this case we assume that the noise-source power is known, and can be represented by $F_{\text{noisesource}}$, which is sometimes also called the excess noise ratio (ENR). This was actually captured at the calibration stage. On combining these two equations we can calculate a factor called $Y_{\text{factor}}$ as

$$Y_{\text{factor}} = \frac{P_{\text{nDUTON}}}{P_{\text{nDUTOFF}}} = \frac{F_{\text{noisesource}} + F_{\text{overall}}}{F_{\text{overall}}} \tag{2.56}$$

$F_{\text{overall}}$ can then be calculated as

$$F_{\text{overall}} = \frac{F_{\text{noisesource}}}{Y_{\text{factor}} - 1} \tag{2.57}$$

As mentioned, $F_{\text{overall}}$ is the overall noise factor, that is, the combination of the DUT noise factor with the spectrum-analyzer noise factor. Fortunately, since the noise factor of the spectrum analyzer is known, the DUT noise factor can be calculated using the noise Friis formula:

$$F_{\text{overall}} = F_{\text{DUT}} + \frac{F_{\text{SA}} - 1}{G_{\text{DUT}}} \tag{2.58}$$

leading to

$$F_{\text{DUT}} = F_{\text{overall}} - \frac{F_{\text{SA}} - 1}{G_{\text{DUT}}} \tag{2.59}$$

The DUT gain can also be calculated using the previous measurement done at the calibration stage, also using the setup of Fig. 2.65. In this case the extra value at the output of the DUT will be measured and compared with the previous noise-source overall power:

$$G_{\text{DUT}} = \frac{P_{\text{nDUTON}} - P_{\text{nDUTOFF}}}{P_{\text{ngeneratorON}} - P_{\text{ngeneratorOFF}}} \tag{2.60}$$

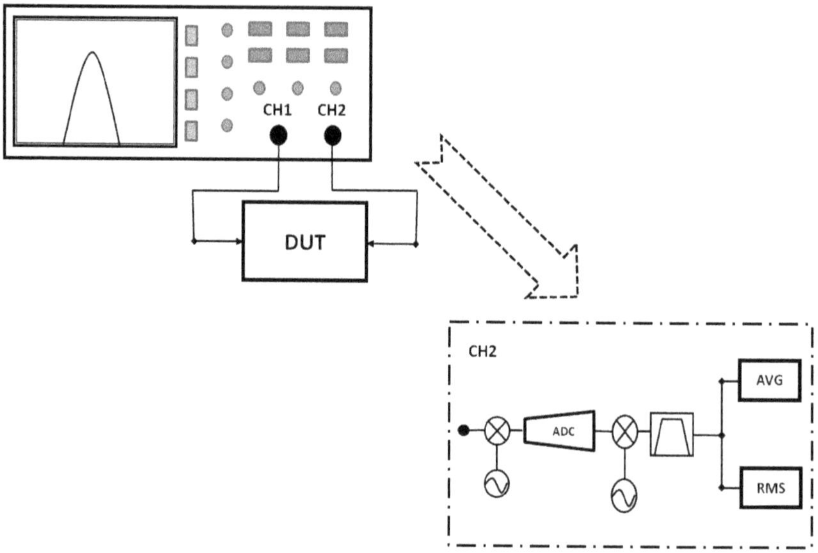


**Figure 2.66** Noise-figure measurement without a noise source.

There are several drawbacks to this method, notably the fact that any error in $F_{\text{noisesource}}$ will appear as an error in the final measurement. Also, the method does not implement any power calibration and does not correct for input mismatch, and therefore the method provides better results for low VSWR values.

## 2.10.2 Noise-figure measurement without a noise source

Another very interesting method for measuring the noise figure of a DUT is completely based on the high quality of instruments nowadays, and on the ability to process a huge amount of data. The method consists of reducing noise by brute force.

The traditional method employed to reduce noise in measured signals is to consider that the noise is a random variable with zero mean and that the signal to be measured is a deterministic signal in a time window. So, if the signal is averaged, its shape is kept constant, since the signal is deterministic, and therefore does not change from measured window to measured window. On the other hand, since noise is random with zero as mean value, the overall value will converge to zero with increasing $O$, with $O$ the number of averages.

Thus, in order to implement this measurement method a simple network analyzer is used, and a single sine wave is measured at the output using two different approaches, as illustrated in Fig. 2.66.

As can be seen from Fig. 2.66, the data is processed using two different mathematical approaches. One approach considers the average (AVG) value of the signal, while the other considers the root-mean-square (RMS) value of the signal. The AVG value will converge to the signal itself, while the RMS value will account for the signal and the noise. The following equations explain how these calculations are done:

$$x_{\text{AVG}} = \frac{1}{O}\sum_{i=1}^{O} x_i \tag{2.61}$$

$$x_{\text{RMS}} = \sqrt{\frac{1}{O}\sum_{i=1}^{O} |x_i|^2} \tag{2.62}$$

On considering now that our signal is corrupted by noise, $x_i = s + n_i$, and applying the two equations, we get

$$|x_{\text{AVG}}|^2 = \left|\frac{1}{O}\sum_{i=1}^{O}(s + n_i)\right|^2 \tag{2.63}$$

$$= |s|^2 + \left|\frac{1}{O}\sum_{i=1}^{O} n_i\right|^2 + \frac{1}{O}\sum_{i=1}^{O}(sn_i^* + s^*n_i)$$

$$x_{\text{RMS}}^2 = \frac{1}{O}\sum_{i=1}^{O} |x_i|^2$$

$$= |s|^2 + \frac{1}{O}\sum_{i=1}^{O} |n_i|^2 + \frac{1}{O}\sum_{i=1}^{O}(sn_i^* + s^*n_i) \tag{2.64}$$

If the difference between the two is calculated, we obtain

$$x_{\text{RMS}}^2 - |x_{\text{AVG}}|^2 = \frac{1}{O}\sum_{i=1}^{O} |n_i|^2 - \left|\frac{1}{O}\sum_{i=1}^{O} n_i\right|^2 \tag{2.65}$$

For large values of $O$ the second term will converge to zero and the final result will contain mainly the measured noise power:

$$P_{\text{noise}} = P_{\text{signal+noise}} - P_{\text{signal}}$$
$$= \frac{1}{RO}\left\{\sum_{i=1}^{O} |n_i|^2 - \frac{1}{O}\left|\sum_{i=1}^{O} n_i\right|^2\right\} \tag{2.66}$$

where $R$ is the system impedance. Using again the well-known $F$ formulas as in Section 1.3.2, we have

$$F_{\text{DUT}} = \frac{\text{SNR}_{\text{input}}}{\text{SNR}_{\text{output}}}$$
$$= \frac{P_{\text{signal}_{\text{input}}}/P_{\text{noise}_{\text{input}}}}{P_{\text{signal}_{\text{output}}}/P_{\text{noise}_{\text{output}}}} \tag{2.67}$$

$P_{\text{signal}_{\text{input}}}/P_{\text{noise}_{\text{input}}}$ can be measured without the DUT being present and following a calibration procedure. It should be stated that, since this measurement is strongly based on power evaluation, a correct calibration is fundamental prior to any measurement, otherwise the overall result will be strongly corrupted by noise. This calibration is supported by a series of attenuators that are connected in sequence at the input/output ports.

In Chapter 4 some examples will be given in order to help the reader understand how a full noise measurement can be done.

### 2.10.3 Accuracy and uncertainty of noise-figure measurement

The accuracy and uncertainty of these noise-figure-measurement instruments are similar to those of spectrum analyzers and vector network analyzers, since they are actually part of the measurement process. Actually, extra errors can appear in these instruments, mainly from the noise source in the approach using noise-source devices. In those cases the noise source should be carefully characterized prior to any measurement. Actually, the operator should be aware that any extra noise sources around the instrument (for instance, mobile phones, WiFi interfaces, fluorescent lights, etc.) can completely degrade the measurement. So we should avoid laboratory spaces where such extra noise sources are present.

Moreover, any mismatch could also degrade your measurement and subsequently the measured $Y$-factor. As presented in other schemes, the use of an isolator can help to minimize the uncertainty problem due to mismatches.

The same problems as those identified with power meters can actually be considered here. For instance, if the measurement is overloaded, nonlinear behaviors can appear and may completely degrade the measured result. So the user should ensure that the input of the power meter (noise-figure analyzer) is not overloading the RF front end.

Uncertainty in this type of instrument can thus be calculated as [20]

$$\begin{aligned}\delta \mathrm{NF}_{\mathrm{DUT}} = \Bigg[ & \left(\frac{F_{\mathrm{overall}}}{F_{\mathrm{DUT}}}\,\delta \mathrm{NF}_{\mathrm{overall}}\right)^2 + \left(\frac{F_{\mathrm{SA}}}{F_{\mathrm{DUT}}G_{\mathrm{DUT}}}\,\delta \mathrm{NF}_{\mathrm{SA}}\right)^2 \\ & + \left(\frac{F_{\mathrm{SA}}-1}{F_{\mathrm{DUT}}G_{\mathrm{DUT}}}\,\delta G_{\mathrm{DUT}_{\mathrm{dB}}}\right)^2 \\ & + P\left(\left(\frac{F_{\mathrm{overall}}}{F_{\mathrm{DUT}}} - \frac{F_{\mathrm{SA}}}{F_{\mathrm{DUT}}G_{\mathrm{DUT}_{\mathrm{dB}}}}\right)\delta \mathrm{ENR}_{\mathrm{dB}}\right)^2\Bigg]^{1/2} \end{aligned} \tag{2.68}$$

where $P$ is 1 for a single frequency, and 0 for a measurement involving frequency conversion. Each of the $\delta$ terms corresponds to the variance of a value.

## Problems

**2.1** If the spectrum occupancy in a mobile network is to be measured, what is the best instrument to use and why?

**2.2** Calculate the sampling bandwidth you should have in a VSA if a WiFi signal with bandwidth 26 MHz is to be measured.

**2.3** Compare an RTSA with a VSA in terms of bandwidth capability and explain the main differences.

**2.4** Consider a CW signal with 1 W of power that needs to be measured in the presence of a 1-μW modulated signal with a power spectral density spread in a bandwidth of 1 MHz. Calculate in this case the minimum dynamic range of an SA.

**2.5** Calculate the optimum sweep time for an SA with RBW 1 Hz and span 1 MHz considering a filter behavior with $k = 1$.

**2.6** What is the increase of the NF in an SA if an attenuator of 10 dB is inserted at the input?

**2.7** What imposes the maximum power limitation in any instrumentation? Focus on power meters and SAs.

**2.8** Explain the main problem that appears in a VNA measurement if the isolation between ports is not zero.

**2.9** What is the main limitation of a VNA if we do not use the through in the calibration process?

**2.10** A comb generator is fundamental in a NVNA. Explain why.

**2.11** Why does one need to do a power calibration in a NVNA configuration?

## References


[1] European Union, Council Directive 89/617/EEC of 27 November 1989 amending Directive 80/181/EEC on the approximation of the laws of the Member States relating to units of measurement.

[2] Agilent Technologies, Agilent fundamentals of RF and microwave power measurements, Application note 64-1C, 2001.

[3] Rohde & Schwarz, Voltage and power measurements – fundamentals, definitions, products (Application note).

[4] D. Humphreys and J. Miall, "Traceable RF peak power measurements for mobile communications," *IEEE Trans. Instrum. Meas.*, vol. 54, no. 2, pp. 680–683, Apr. 2005.

[5] H. Gomes, A. R. Testera, N. B. Carvalho, M. Fernandez-Barciela, and K. A. Remley, "Diode power probe measurements of wireless signals," *IEEE Trans. Microwave Theory Tech.*, vol. 59, no. 4, pp. 987–997, Apr. 2011.

[6] M. B. Steer, *Microwave and RF Design: A Systems Approach*. Herndon, VA: SciTech Publishing, 2010.

[7] J. C. Pedro and N. B. Carvalho, *Intermodulation Distortion in Microwave and Wireless Circuits*. New York: Artech House, 2003.

[8] C. Brown, *Spectrum Analysis Basics*. Hewlett-Packard Company, 1997.

[9] C. Rauscher, *Fundamentals of Spectrum Analysis*. Rohde & Schwarz, 2001.

[10] P. S. R. Diniz, E. A. B. da Silva, and S. L. Netto, *Digital Signal Processing – System Analysis and Design*. Cambridge: Cambridge University Press, 2010.

[11] G. F. Engen and C. A. Hoer, "Thru-reflect-line: An improved technique for calibrating the dual six-port automatic network analyzer," *IEEE Trans. Microwave Theory Tech.*, vol. 27, no. 12, pp. 987–993, Dec. 1979.

[12] A. Lewandowski, D. F. Williams, P. D. Hale, J. C. M. Wang, and A. Dienstfrey, "Covariance-based vector-network-analyzer uncertainty analysis for time- and frequency-domain measurements," *IEEE Trans. Microwave Theory Tech.*, vol. 58, no. 7, pp. 1877–1886, Jul. 2010.

[13] BIPM, Evaluation of measurement data – guide to the expression of uncertainty in measurement, http://www.bipm.org/en/publications/guides/gum.html, 2008.

[14] D. Schreurs, "Applications of vector non-linear microwave measurements," *IET J. Microwaves, Antennas Propagation*, vol. 4, no. 4, pp. 421–425, Apr. 2010.

[15] P. Blockley, D. Gunyan, and J. B. Scott, "Mixer-based, vector-corrected, vector signal/network analyzer offering 300 kHz–20 GHz bandwidth and traceable phase response," in *IEEE MTT-S International Microwave Symposium*, Jun. 2005, pp. 1497–1500.

[16] T. Van den Broeck and J. Verspecht, "Calibrated vectorial nonlinear-network analyzers," in *IEEE MTT-S International Microwave Symposium*, May 1994, pp. 1069–1072.

[17] T. S. Clement, P. D. Hale, D. F. Williams *et al.*, "Calibration of sampling oscilloscopes with high-speed photodiodes," *IEEE Trans. Microwave Theory Tech.*, vol. 54, no. 8, pp. 3173–3181, Aug. 2006.

[18] D. Humphreys, M. Harper, J. Miall, and D. Schreurs, "Characterization and behavior of comb phase-standards," in *European Microwave Conference (EuMC)*, Oct. 2011, pp. 926–929.

[19] P. D. Hale, C. M. Wang, D. F. Williams, K. A. Remley, and J. Wepman, "Compensation of random and systematic timing errors in sampling oscilloscopes," *IEEE Trans. Instrum. Meas.*, vol. 55, no. 6, pp. 2146–2154, Dec. 2006.

[20] Agilent Technologies, Noise figure measurement accuracy – the *y*-factor method, Application note 57-2, 2010.

# 3 Signal excitation

## 3.1 Introduction

This chapter is devoted to signal excitation for RF characterization. In that respect the main types of excitation used in the microwave field will be presented and their capabilities for wireless-system characterization, identification, and modeling will be explained. The main focus of this chapter is on the generation of special signal patterns and the explanation of the capabilities that those signals bring to the wireless metrology. This approach will complement previous chapters, by explaining how to generate and how to gather important information from specially designed signals. The chapter will start with the one-tone or single-sinusoid excitation.

In linear systems, figures of merit spanning power gain, noise figure, VSWR, bandwidth, etc. are mainly measured and obtained using a single sinusoid. Since in this type of system superposition laws are valid, the extension of those figures of merit to other forms of excitation that are different from the ones which were used for their identification is valid. So, normally, linear systems' figures of merit are measured using one tone, and then extrapolated very easily to other forms of excitation.

In nonlinear systems this extrapolation is not so simple, and in certain cases it is not even possible, since a nonlinear system does not obey superposition and thus the one-tone excitation is not enough to gather all the characteristics of the nonlinear system. This fact will give rise to the next section, and the chapter will evolve to consider more complex signals such as the two-tone (two-sinusoid) signal, which is a signal that is typically used for evaluation of nonlinear distortion, mainly intermodulation distortion.

The chapter will then move forward to multi-sine excitation, which is very important, not only from a spectral point of view, but also from a time-domain point of view, since some special care should be taken in order to generate multi-sine excitations for testings of real telecommunications RF components.

We will then progress to a discussion of complex modulated signals, where the concepts behind arbitrary waveform generators will be presented and discussed. Finally, some more dedicated and specially designed signals for component modeling will be approached by presenting the generation and evaluations of chirp signals.

## 3.2 One-tone excitation

One-tone excitation and single sinusoids are actually the orthogonal basis for evaluation of Fourier series. This is why the single sinusoid is the preferred form of excitation when we are concerned with measuring steady-state and spectrum contents. For instance, linear systems are well characterized by using single sinusoids as the excitation. The DUT is then measured by accounting for its output amplitude and phase deviation in the frequency domain. This approach actually allows us to measure the complete frequency-domain transfer function $H(j\omega)$. The input excitation in a one-tone excitation is given by

$$x(t) = A_{\mathrm{i}} \cos(\omega t + \theta_{\mathrm{i}}) = \mathrm{Re}\left[A_{\mathrm{i}} e^{-j\theta_{\mathrm{i}}} e^{-j\omega t}\right] \tag{3.1}$$

where $A_{\mathrm{i}}$ is the input amplitude and $\omega = 2\pi f$ is the fundamental radian frequency.

In linear systems the output of the evaluated component is measured at the same input frequency, which will be referred to from now on as the fundamental frequency. Since the linear system affects only the phase and amplitude of the input signal, the output will be

$$y(t) = A_{\mathrm{o}}(\omega)\cos(\omega t + \theta_{\mathrm{o}}(\omega)) = \mathrm{Re}\left[A_{\mathrm{o}}(\omega) e^{-j\theta_{\mathrm{o}}(\omega)} e^{-j\omega t}\right] \tag{3.2}$$

From Eq. (3.2) it can be seen that the output will have a change in amplitude and in phase, and thus it can perfectly characterize the input–output relationship and the linear transfer function $H(j\omega)$, as expressed by

$$\begin{aligned} H(j\omega) &= \frac{\tilde{A}_{\mathrm{o}}(\omega)}{\tilde{A}_{\mathrm{i}}(\omega)} \\ &= \frac{a_{\mathrm{or}} + j a_{\mathrm{oi}}}{a_{\mathrm{ir}} + j a_{\mathrm{ii}}} \end{aligned} \tag{3.3}$$

where $a_{\mathrm{or}}$ and $a_{\mathrm{ir}}$ are the real output and input components, and $a_{\mathrm{oi}}$ and $a_{\mathrm{ii}}$ are the imaginary output and input components, respectively.

This function can then be plotted using a phase and amplitude diagram, and thus it characterizes completely the DUT.

In microwave circuits and systems, instead of representing the signals as $x(t)$, it is usual to represent the incident and scattered waves into each device port, as explained in Section 1.2.1. In that case, for instance in an $i$-port device, the measured quantities will be the incident wave in port $i$, $\tilde{a}_i$, and the scattered wave at the same port, $\tilde{b}_i$.

Thus any of the characterization figures of merit for linear RF devices, explained in Section 1.2.1 can be characterized perfectly using this type of excitation. Moreover, if the amplitude of the input excitation is varied, it is also possible to measure nonlinear figures of merit with this excitation, such as the AM–AM and AM–PM figures introduced in Section 1.5.

### 3.2.1 One-tone generation mechanisms

In order to generate a single sinusoid, several approaches can be followed, from simple RF oscillators to the more complex digital versions like the direct digital synthesis (DDS) oscillator. In any case it is important to consider that the oscillator signal should be as close as possible to a frequency-domain impulse at a certain frequency. Figure 3.1 presents an ideal sinusoidal excitation, in both domains (time and frequency).

It is very important to consider several parameters for evaluating the quality of an RF sinusoid. The parameters of interest include the phase noise, output power, and frequency stability.

As was explained in Section 1.10.1.2, phase noise is in fact a very important and difficult parameter to control, and a low value is fundamental for good RF system performance.

The topology of a typical RF source is presented in Fig. 3.2. An RF source is built in three main blocks, namely the RF reference block, the synthesizer part, and the output section. These blocks will now be discussed individually. Note that here we will show

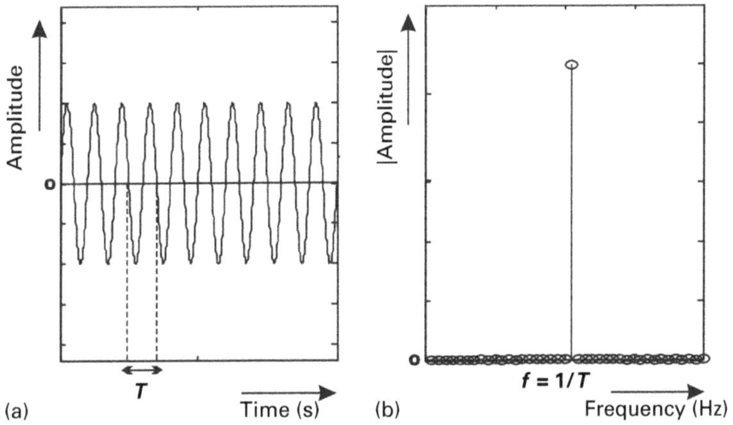


**Figure 3.1** An ideal RF single sinusoid: (a) in the time domain and (b) in the frequency domain.

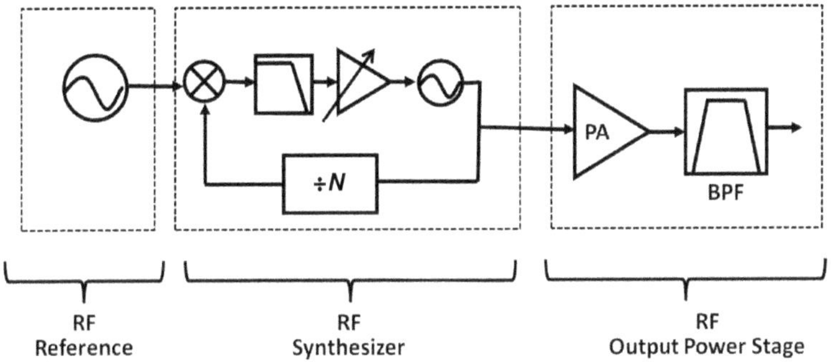


**Figure 3.2** The three RF generator blocks, namely the RF reference, the RF synthesizer and the RF output power stage.

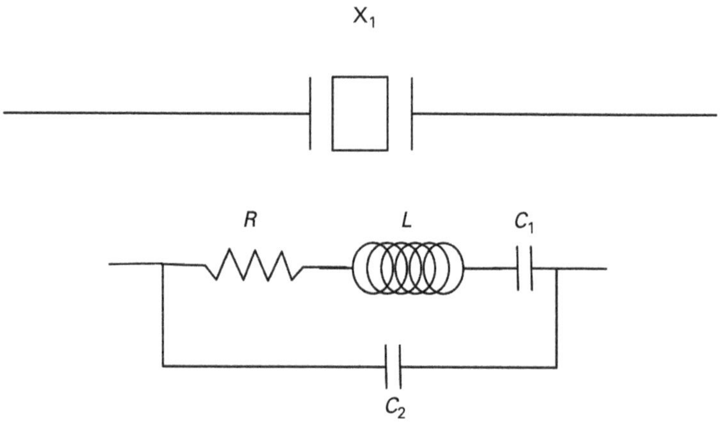


**Figure 3.3** A typical reference crystal oscillator. In the upper part the symbol used is displayed. In the bottom part the model of a crystal oscillator is sketched. This is composed of an *RLC* circuit in parallel with a capacitor. The *RLC* circuit imposes a resonance frequency of the crystal oscillator.

and identify approaches for the construction of RF generators; how to actually design RF oscillators is beyond the scope of this book. Several good references can be found in this field, such as [1, 2].

### 3.2.1.1 The reference signal

The reference signal can be provided by any good oscillator. Oscillators are most commonly realized using quartz, typically as crystal oscillators. Nevertheless, these crystalline oscillators are affected by such parameters as aging, temperature, and line voltage. Temperature variation actually causes the most severe form of signal-quality degradation, and thus, in order to improve its characteristics, the crystal oscillator can be further compensated for temperature variation, by using a thermally compensated oscillator (TCXO) or by putting it in a controlled environment such as an oven, being in that case an oven-controlled oscillator (OCXO). Figure 3.3 presents a crystal oscillator.

It should be noted that external references could also be used, which can improve or sometimes degrade the overall reference. These reference oscillators most commonly work at 10 MHz in laboratory environments.

### 3.2.1.2 The synthesizer block

The output frequency of the generator derived from the reference oscillator is created by having recourse to a synthesizer. In the majority of cases the synthesizer is based on the well-known principle of a phase-locked loop (PLL) [3]. The use of PLLs is in fact a clever way to create a synthesized sinusoidal oscillator. In this case the frequency can be selected using a digital keyword, typically by using digital frequency dividers that can be introduced into the control loop.

Figure 3.4 presents one possible configuration, where we see the reference oscillator, mixer (called a phase detector), filter, VCO, and frequency divider. The basic principle

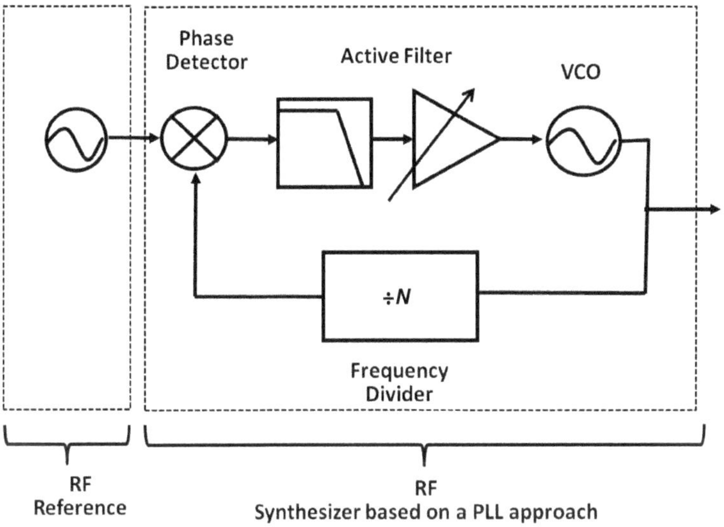


**Figure 3.4** The ideal PLL approach, including the RF reference and the RF synthesizer based on a PLL configuration.

is based on the fact that, when the reference oscillator is mixed with the feedback signal, it will create an output signal that will be filtered out by the low-pass filter before being fed to the VCO. The low-frequency voltage at the input of the VCO will make the VCO run at an RF frequency that will be further subdivided by the loop frequency divider. This process will then continue until the two signals at the mixer run at the same frequency and in phase, which means that the output of the mixer will be, after the loop has converged, a DC voltage. This voltage will then feed the VCO, and thus the output will be maintained at a constant frequency. When the two frequencies at the mixer are equal, we say that the PLL is locked. The filter bandwidth should be chosen in order to filter out any high value of frequency, but not to be as narrow as possible, since in that case the PLL will not converge and will not lock for a high separation between frequencies. The frequency will then be decided using the loop frequency divider, where we can choose a correct value for generating the RF frequency.

For instance, consider a crystal reference oscillator running at 10 MHz, and a 2.4-GHz output signal as the main RF design objective. The frequency division should be at least $2400/10 = 240$. This division number is usually selected using a digital keyword that is fed to the digital divider. More information and design rules for PLL synthesizers can be found in [3].

If the reader carefully looks at the PLL structure, it is possible to observe that one of the main components is a voltage-controlled oscillator (VCO). The VCO is in fact the core component in the PLL, and it is responsible for the overall generation mechanism. It can be built from a simple oscillator such as one in the Colpitts and/or Hartley configuration [1], both of which configurations are illustrated in Fig. 3.5. In these circuits the oscillator frequency is usually controlled by changing the feedback loop using a capacitance that changes with voltage, such as a varicap.

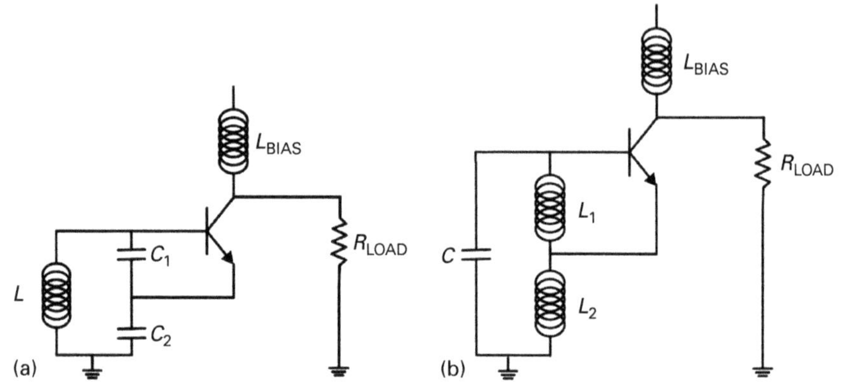


**Figure 3.5** Possible solutions for VCO implementation: (a) the Colpitts configuration and (b) the Hartley configuration.

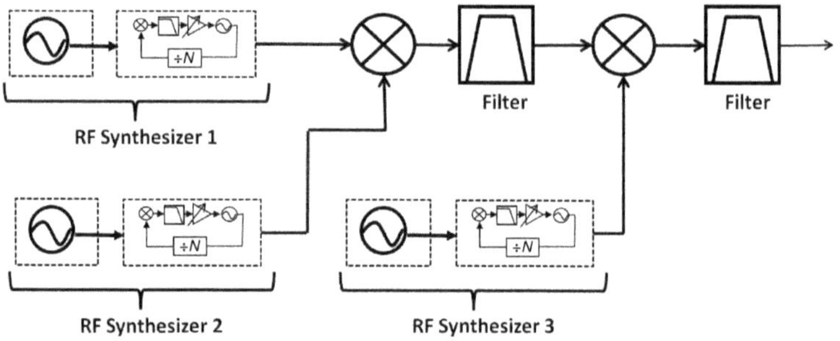


**Figure 3.6** The DFS approach. In this case three RF synthesizers are used in combination to create a higher output frequency.

In these VCO configurations special care should be taken to guarantee a good phase-noise pattern. In fact the oscillator is controlled by the feedback loop of the configuration, and it can be optimized by using some kind of resonator. More information can be obtained from [1].

Nevertheless, PLLs are not suitable for very high frequencies in the microwave or millimeter-wave range. In that case the generator can be built using a synthesized version for the lower-frequency part, and then up-converted to higher frequencies by using a cascaded mixing strategy, as presented in Fig. 3.6. This procedure is called direct frequency synthesis (DFS). It allows the generation of an output frequency whereby several RF oscillators are mixed and filtered continuously in order to create an output signal at very high frequency. Sometimes a YIG (yttrium iron garnet) oscillator is used to improve the RF generation. Such a YIG oscillator is actually tuned with magnetic fields.

There are also some approaches that are implemented completely in the digital domain, called direct digital synthesis (DDS). The principle is shown in Fig. 3.7. In that case a digital form of the sine-wave signal is first uploaded by software to a memory

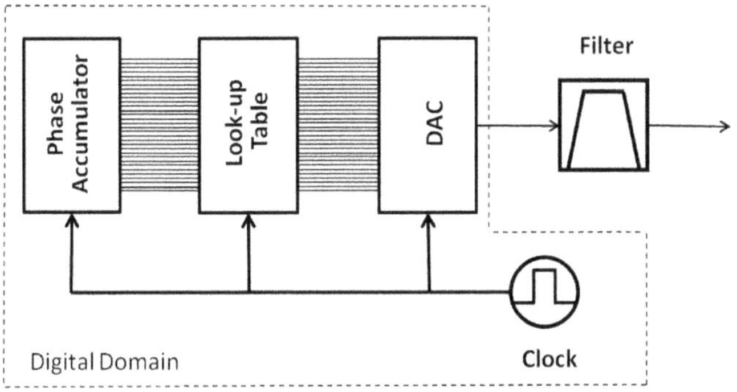


**Figure 3.7** The DDS approach. The diagram shows the digital part of this approach, including a phase accumulator, a look-up table, and the final-stage DAC.

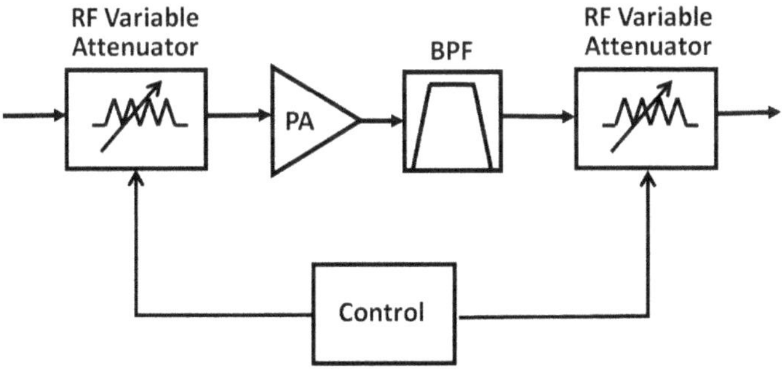


**Figure 3.8** The output stage of the RF generator.

look-up table. This table is then read by using a phase accumulator that will traverse all the table points with a predetermined frequency. The read values will then be fed into a digital-to-analog converter (DAC). Thus the frequency in this case will be proportional to the speed at which the values in the look-up table are traversed. At the output we need a bandpass filter to guarantee that the sine wave does not resemble a DC staircase form, but is really a sine wave. These generators are mainly used in the low-frequency range. For example, nowadays commercially available DDS generators operate at up to 3 GHz.

### 3.2.1.3 The output stage

The generator is finally composed of an output block, that contains the circuitry that is needed in order to level up the signal, and to control its amplitude. The scheme is presented in Fig. 3.8.

This block is the main cause of non-ideal responses from the generator, since the output amplifier can operate in the nonlinear regime, and in that case it will generate several harmonic contents and spurious signals that will degrade the overall signal purity. Actually, the procedures and evaluation that were presented in Section 1.8 in relation to power amplifiers can be applied directly here. This means that most RF generators

**Table 3.1** Datasheet frequency parameters for a sinusoidal-signal generator

| | |
|---|---|
| Frequency | |
| Range (model) | |
| Model 1 | 10 MHz to 20 GHz |
| Model 2 | 10 MHz to 30 GHz |
| Model 3 | 10 MHz to 40 GHz |
| Resolution | 0.1 Hz |
| Setting time | <(11 ms + 4 ms/GHz) |
| Phase offset | Adjustable in 2° steps |
| Reference frequency | |
| Aging | $1 \times 10^{-6}$/year |
| Temperature impact | $2 \times 10^{-6}$ |
| Warm-up time | 10 min |
| Output for internal reference | |
| Frequency | 10 MHz |
| Level $V_{rms}$ | 1V |
| Source impedance | 50 Ω |
| Input for internal reference | |
| Frequency | 1 MHz to 20 MHz |
| Input level $V_{rms}$ | 0.1 V to 5 V |
| Input impedance | 200 Ω |

should be operated at a value smaller than the maximum available, since at the maximum the harmonic content can alter the measurements being made. Most modern RF generators present an "uncal" warning on reaching these points of operation.

### 3.2.2 One-tone instrumentation

In the previous section some of the procedures for generating sinusoidal signals were presented. Nevertheless, it is important to consider several figures of merit when selecting an RF generator. Tables 3.1 and 3.2 present typical datasheets of an RF generator.

As can be seen from the datasheets, several figures of merit are considered, namely the frequency of operation, phase-noise level, maximum output power, frequency step, attenuation step, etc. Let us discuss these parameters individually.

#### 3.2.2.1 Frequency parameters

The RF parameters concern the frequency of operation. For the example considered, the frequency range of the instrument spans from a few MHz to several GHz. The frequency resolution is the next parameter, where, e.g., 0.1 Hz is given as the resolution. The resolution is defined as the minimum difference in frequency between two sinusoids which still allows one to have two distinct peaks in the spectrum.

**Table 3.2** Sinusoidal generator datasheet power parameters

| | | |
|---|---|---|
| Level | | |
| Frequency range | Maximum level | |
| 10 MHz to <2 GHz | >+17 dBm | |
| 2 GHz to 20 GHz | >+11 dBm | |
| Accuracy | | |
| Frequency range | Level | Accuracy |
| 10 MHz to <2 GHz | >+10 dBm | <±1.2 dB |
| | >−10 dBm | <±0.6 dB |
| | >−60 dBm | <±0.9 dB |
| | ≤−60 dBm | <±1.4 dB |
| 2 GHz to 20 GHz | >+10 dBm | <±1.3 dB |
| | >−10 dBm | <±0.7 dB |
| | >−60 dBm | <±1.0 dB |
| | ≤−60 dBm | <±1.5 dB |
| <20 GHz to 40 GHz | >+10 dBm | <±1.5 dB |
| | >−10 dBm | <±0.9 dB |
| | >−60 dBm | <±1.2 dB |
| | ≤−60 dBm | <±1.7 dB |
| Output impedance | 50 Ω | |
| VSWR | | |
| $f \leq 20$ GHz | <2 | |
| $f > 20$ GHz | <2.2 | |
| Settling time | <20 ms | |
| Attenuator setting range | 0 dB to 20 dB | |

Another important parameter is the settling time, which presents important information about the speed at which the RF generator can change from one frequency to another. In this case it is on the order of milliseconds. This specification is very important when using the RF generator in combination with a network analyzer, for instance.

It is important to notice that different frequency bands are identified for each parameter. This is due to the fact that, to achieve such a broad bandwidth, namely from MHz to GHz, the generator will actually need to be run with different configurations, and thus with different specifications.

The accuracy of the oscillator is further given by the change with temperature (specified in Hz per year or per month), change in line voltage (specified as a percentage of voltage change), and change in calibration.

Other parameters are also provided, such as the external reference signal to be used if necessary. The reference signal is usually a signal at frequency 10 MHz, which is the standard in laboratory instrumentation. Depending on the options of the generator, other information may be available, such as the sweeping points for step generators or the sweeping time for ramp generators. It should be mentioned that sometimes these

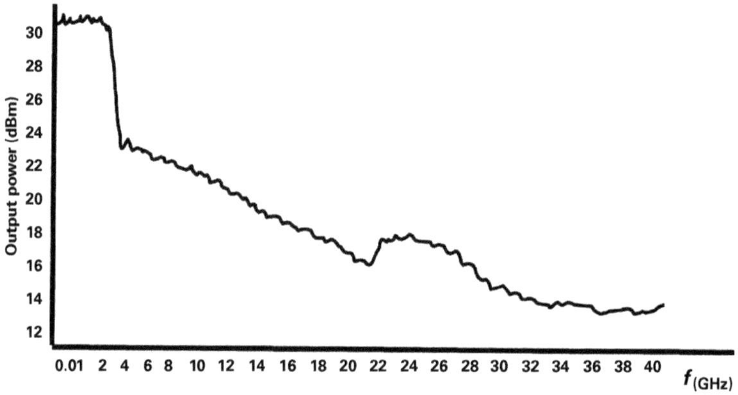


**Figure 3.9** An example of output power versus frequency of an RF generator.

generators accept also triggers that allow them to be connected with other instrumentation in order to measure quantities in a synchronized manner.

### 3.2.2.2 Output power parameters

The next main group of parameters is related to the output power, Table 3.2. In this case it is important to know the values of the minimum stable output power and the maximum output power, and moreover to understand that the output power can change with the frequency band due to the fabrication of the generator. In this case the step in power is also very important, because, for instance for measuring the AM–AM figure of merit, the power step will impose the achievable resolution of the measurement. Figure 3.9 presents an example of the variation of the maximum output power with frequency.

It is also important to take account of the amplitude switching speed. This is mainly important for power measurements, since the sweeping time can change the operation of the device in certain contexts, due to the DUT dynamics. As before, the accuracy is described for each frequency band (expressed in dB), as well as the resolution in amplitude and the stability with temperature. Another important parameter is the output impedance, which should be 50 Ω for standard RF laboratory instrumentation. Since 50 Ω is not obtained in all frequency bands, the VSWR (as presented in Section 1.3.1.1) is given for each band. We should be aware that the VSWR value should typically stay below 2 for correct operation of the system.

### 3.2.2.3 Spurious-signal generation parameters

Finally let us describe some other parameters that represent non-ideal behaviors of the RF generator, and that can completely degrade the overall measurement. These parameters are mainly related to the generation of spurious signals, called spectral purity, which stands for the creation of harmonics or other frequency components as well as the generation of phase noise.

The first main problem is related to the generation of harmonics, which will completely degrade the identification of nonlinear phenomena in the measured DUT. That

**Table 3.3** Datasheet information on to spurious-signal generation

| | | | | |
|---|---|---|---|---|
| Harmonics | | | | |
| $f < 1.8$ GHz | <−30 dBc (<+8 dBm) | | | |
| $f \geq 1.8$ GHz | <−40 dBc (<+10 dBm) | | | |
| Subharmonics | | | | |
| $f \leq 20$ GHz | <−35 dBc | | | |
| $f > 20$ GHz | <−40 dBc | | | |
| Non-Harmonics at >10 kHz from carrier | | | | |
| $f < 2$ GHz | <−60 dBc | | | |
| 2 GHz to 20 GHz | <−60 dBc | | | |
| $f$>20 GHz | <−54 dBc | | | |
| | SSB phase noise, 1 Hz bandwidth | | | |
| | Offset from carrier | | | |
| Frequency range | 100 Hz | 1 kHz | 10 kHz | 100 kHz |
| 10 MHz to <2 GHz | <−64 dBc | <−93 dBc | <−104 dBc | <−104 dBc |
| 2 GHz to 10 GHz | <−64 dBc | <−93 dBc | <−105 dBc | <−105 dBc |
| >10 GHz to 20 GHz | <−58 dBc | <−87 dBc | <−99 dBc | <−99 dBc |

is why, as will be seen in Chapter 4, a low-pass filter is sometimes used at the output of the CW generator in order to reduce the harmonic content. Manufacturers present the information on this harmonic generation for different frequency bands and for a certain output power in their datasheets, as in Table 3.3.

The reader should be aware that, if the output power changes, the harmonics will also change accordingly, due to the fact that the harmonics are intrinsically a nonlinear phenomenon of the output amplifier stage. It is also important to take care of the subharmonic generation and the spurious signals appearing at other non-harmonic frequencies. These are mainly generated due to frequency mixing inside the generator between the main oscillator and oscillator references or reference multiples.

It should also be stressed that broadband noise is a very important parameter when generating low-power signals that are close to the noise level.

Finally other information and parameters are presented from a general point of view, such as power consumption, acoustic noise, etc. Figure 3.10 presents typical RF generators from several vendors.

## 3.3 Two-tone excitation

One-tone characterization is in fact the most widely used type of signal excitation in RF laboratories. If we are concerned with the characterization of devices that are linear, then a single tone suffices to obtain and extract most of the component information. Strictly

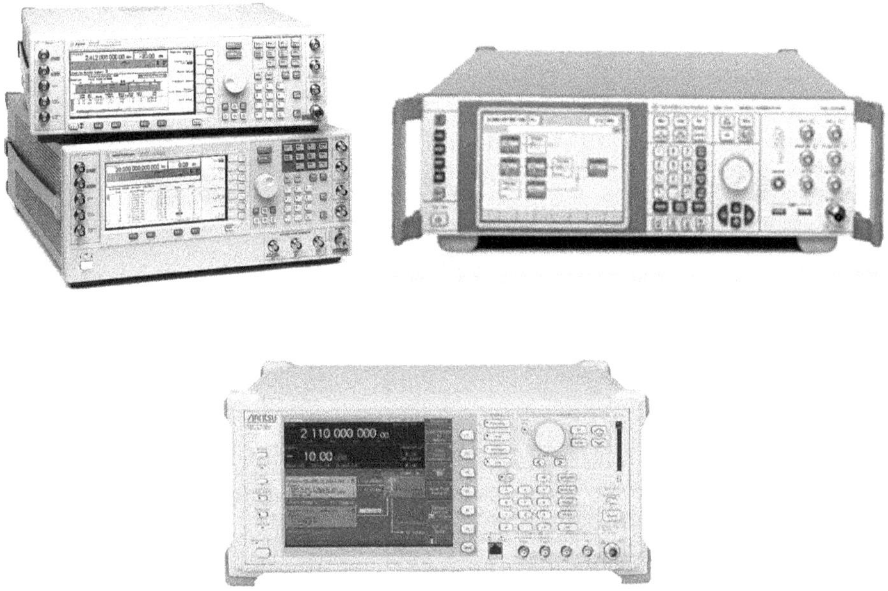

**Figure 3.10** RF generators from (a) Agilent Technologies, (b) Rohde & Schwarz, and (c) Anritsu. © Agilent Technologies, Rohde & Schwarz, and Anritsu.

speaking, there are some linear systems that do not follow this rule [4], but most of the time they are not excited by RF signals.

Nevertheless if nonlinear devices are to be characterized, then one tone is no longer an optimum choice, and a better input signal should be used. For instance an improved approximation is provided by two-tone excitation, since it mimics the real signals that the DUT will encounter in the real world. In this case, as was seen previously in Section 1.4.1, it will also allow the generation of in-band distortion and signal bandwidth identification.

Actually, a nonlinear DUT excited by a two-tone signal is able to generate harmonics of the input signal as well as other newly generated components that appear close to the fundamental frequencies, close to the harmonics (called spectral regrowth), and close to DC (called baseband components). As in Section 1.4.1, the in-band signals are very important since they allow us to observe the nonlinear distortion generated by the DUT, and they are the key components for nonlinear distortion in bandpass systems. The baseband signals allow us to excite the well-known long-term memory effects, which are not excited by single-tone excitation.

The two-tone excitation is nothing more than a summation of two sinusoids:

$$x(t) = A_{\mathrm{i}1}\cos(\omega_1 t) + A_{\mathrm{i}2}\cos(\omega_2 t) \tag{3.4}$$

The output of a nonlinear device will then be given by

$$y_{\mathrm{NL}}(t) = \sum_{h=1}^{\infty} A_{\mathrm{o}h}\cos(\omega_{\mathrm{o}h} t + \theta_{\mathrm{o}h}) \tag{3.5}$$

where $\omega_o h = m\omega_1 + n\omega_2$ and $m, n \in \mathbb{Z}$.

The output is thus composed of a very large number of newly generated tones, at $m\omega_1 + n\omega_2$.

### 3.3.1 Two-tone generation mechanisms

For two-tone generation, the obvious and straightforward mechanism is to use two single-sinusoid generators and combine them in a single transmission line or RF cable. This scheme is illustrated in Fig. 3.11. Thus, if each of the generators generates a different frequency, the overall generation will be composed of a two-tone signal.

In this simple setup two single-tone generators are used to generate each of the tones of the input tone excitation. It is important to note that each of them is filtered and passes through an isolator. The filter prevents any harmonic of the generator interfering with the measurement, and the isolator is used in order to minimize any signal that could possibly be reflected from the power combiner and subsequently mix with the generated frequency at the generator output front end. This is in fact very important in two-tone generators, since if this happens then the overall measurement can be completely corrupted. The two signals are then combined in the power combiner, creating the final two-tone excitation.

Another type of two-tone generation mechanism is the use of an arbitrary-waveform generator (AWG). This type of generator is similar to the DDS (see Section 3.2.1), but here the two-tone signal is first uploaded to a look-up table, and then, using a DAC, the $I/Q$ signals are generated at baseband, and can then be further up-converted to the RF frequency of interest. Figure 3.12 presents this type of configuration.

Concerning this generator, the reader should be aware of several problems that could degrade the correct measurement of nonlinear effects originating from the DUT. In order to understand these problems, we should consider the overall generation mechanism.

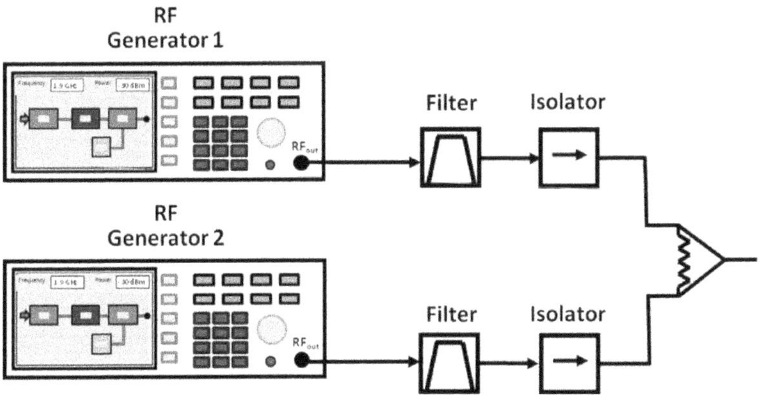


**Figure 3.11** A two-tone generator using single CW generators.

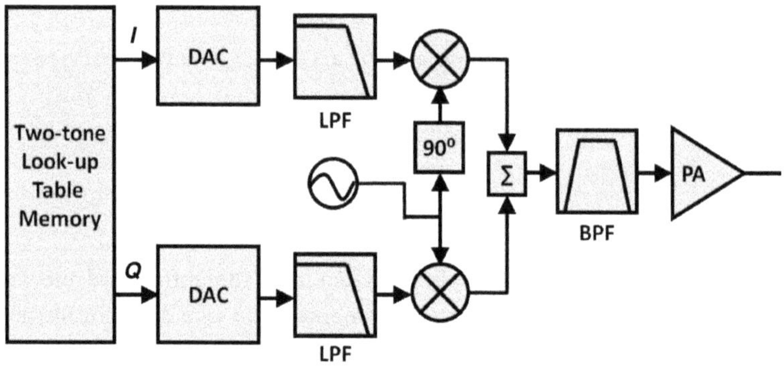


**Figure 3.12** The internal architecture of an arbitrary-waveform generator using an $I/Q$ modulator stage followed by a PA.

First the baseband signal is generated in the digital domain, represented by

$$\tilde{x}(t) = x_I(t) + jx_Q(t) \tag{3.6}$$

This signal is further converted to an analog version by using a DAC as shown in Fig. 3.12, and then filtered out. Assuming a good linear behavior of the DAC component, the signal is then up-converted to RF by using an $I/Q$ modulator, and if necessary an extra up-converter stage. In the two-tone case, one possibility would be to generate a single tone at baseband, and then up-convert it to RF, generating a two-tone signal.

The baseband in this case can be represented by

$$\begin{aligned}\tilde{x}(t) &= x_{\mathrm{I}}(t) + jx_{\mathrm{Q}}(t) \\ &= Ae^{j(\delta\omega_{\mathrm{BB}}t+\theta_{\mathrm{BB}})} + Ae^{-j(\delta\omega_{\mathrm{BB}}t+\theta_{\mathrm{BB}})}\end{aligned} \tag{3.7}$$

where $x_{\mathrm{I}}(t)$ and $x_{\mathrm{Q}}(t)$ are the in-phase and quadrature signals over time, $A$ is the tone's amplitude, $\delta\omega_{\mathrm{BB}}$ is the radial frequency of the baseband signal, sometimes also called the video bandwidth, and $\theta_{\mathrm{BB}}$ is the tone's phase:

$$\begin{aligned}x_{\mathrm{RF}}(t) &= \mathrm{Re}\left(\tilde{x}(t)e^{j\omega_{\mathrm{LO}}t}\right) \\ &= x_{\mathrm{I}}(t)\cos(\omega_{\mathrm{LO}}t) - x_{\mathrm{Q}}(t)\sin(\omega_{\mathrm{LO}}t) \\ &= \frac{A}{2}[\cos((\omega_{\mathrm{LO}} - \omega_{\mathrm{BB}})t - \theta_{\mathrm{BB}}) + \cos((\omega_{\mathrm{LO}} + \omega_{\mathrm{BB}})t + \theta_{\mathrm{BB}})]\end{aligned} \tag{3.8}$$

As can be seen from Eq. (3.8), the output signal will have components at $\omega_{\mathrm{LO}} - \omega_{\mathrm{BB}}$ and $\omega_{\mathrm{LO}} + \omega_{\mathrm{BB}}$. In this case $x_{\mathrm{Q}} = 0$, since we use the same phase for each tone, but, if different phases are used, then $x_{\mathrm{Q}} \neq 4$.

Here $\omega_{LO}$ is exactly in the middle of the two-tone generation, and due to the fact that some DC component appears at the DAC output and also due to a non-ideal isolation, it will appear at the output as a spurious signal that sometimes can degrade the overall measurement. Algorithm 3.1 presents the MATLAB code to create such a signal, and Fig. 3.13 presents these results.

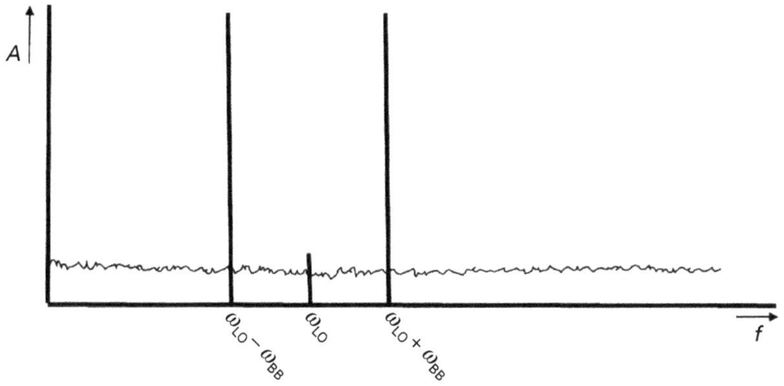


**Figure 3.13** The output of an arbitrary-waveform generator, where the LO leakage is also visible.

ALGORITHM 3.1 (MATLAB code for two-tone baseband generation)

```
% Constructing waveform data. The I–Q data is in the form of a vector that contains
a series of complex numbers (the form is I + jQ).
fs = 40e6; % the sampling frequency 40 MHz
ts = 1/fs; % the sampling time
df = 100e3; % the frequency resolution 100 kHz
T = 200/df; % the number of periods to generate 200 periods
t = ts:ts:T; % construction of the time-domain vector
n = 2; % number of tones, always even, in this case 2
delf = 1e6; % separation between tones 1 MHz
x = 0;
for aa = 1:n
    fx1 = ((aa – 1)*delf) – (delf/2); % generates each tone frequency
    x = x + 1*exp(j*(2*pi*fx1*t)); % complex envelope values
end
% then send the information to the AWG
```

As can be seen from the algorithm, the envelope signal is centered around DC and it is in a complex, $I + jQ$, format. Most of the instrumentation used actually imposes that the carrier frequency is then passed to the instrument itself. On looking at Fig. 3.12, one sees that what has actually been created with this algorithm is only the $I$ and $Q$ information.

In Fig. 3.13 a spurious signal can be seen in between the two sinusoids. This spurious signal is due to the leakage of the local oscillator. One way to obviate this is to use a baseband signal that is already slightly offset in terms of the center frequency, so that the output will appear as a four-tone signal, where two tones are copied at each side of the local oscillator. Again it should be stressed that a correct filtering of the local oscillator and, in this case, the image frequency should be done before applying the generated two-tone excitation to the DUT. One will then encounter the same problems as in superheterodyne configurations [5], where the image frequency could not be too close to the LO frequency since it is necessary to be able to filter it out, and it could also not be too far from the LO frequency due to the need for very-wideband circuitry for the baseband signal, as was also seen in Section 2.3.2.

Moreover, in this type of generator another problem arises if we want to create an output signal with high output power, since the two tones will traverse the output power amplifier, as seen in Fig. 3.12, and thus distortion can arise, creating a huge number of spurious spectral components. In this case an approach that can be used to minimize this nonlinear distortion-generation mechanism is to generate the two-tone signal at a lower power, and use an external linear power amplifier with a high $IP_3$. Nevertheless, this solution should be used carefully due to the possibility of a high noise pattern being inserted by the external power amplifier, so a compromise should be made between spurious output distorted signals and a rise in the noise level. In Section 4.2.3 some examples of how to use this approach are given.

## 3.4 Digitally modulated signals

Digitally modulated signals are in fact the most important type of excitation which new and future wireless systems need for correct evaluation of the figures of merit that have significance in real operations. In this respect we can divide modulated signal generation into two main areas, namely the signals needed for characterization and those for modeling of RF components.

In the first case the signals are perfectly generated using some kind of algorithm at the baseband, and then up-converted conveniently to RF using a scheme similar to what was presented in Fig. 3.12. More information about this is presented in Section 3.4.2. The other specially designed modulated signals are normally used for modeling purposes, and in that case the most important problems to be resolved concern the quality of the signal in terms of the state space to be covered, and hence the capability to identify the highest number of different system states, combined with the need for good periodic signals in order to improve the quality of measurements.

It is well known that the best test signal for nonlinear modeling and characterization is the one that exactly matches the real signal which will be the input to the DUT in real operation. Following this strategy for better excitation signals (better in the sense that they are able to mimic as closely as possible the real telecommunication signals), several excitations with continuous spectra can be used. However, such signals are not always available, so some alternatives could be used, one of which is narrowband Gaussian noise (NBGN) [6]. Unfortunately, signals (real ones) that are completely random and thus with continuous spectra are very difficult to measure using ordinary instrumentation, and moreover very difficult to generate in a systematic measurement.

That is one of the reasons why some other forms of signal are gaining importance in the instrumentation field, since they can be generated very efficiently, and can be recorded for systematic measurements and thus modeling. One of those types is the multi-sine signal, which will be discussed in detail next.

### 3.4.1 The multi-sine

As was seen in the description of the instrumentation in Chapter 2, the use of periodic signal excitation is fundamental for correct gathering of the signal information and thus assessment of the impact of the system on the excitation signal, which is one of the reasons why multi-sine excitation has become of great importance in metrology for RF modeling. Moreover, some instruments cannot handle signals with continuous spectra, in the case of LSNAs or NVNAs, and in that case using multi-sines is the obvious solution for the measurement bench.

The multi-sine signal consists of the sum of several sines (tones). Equation (3.9) presents a typical multi-sine signal:

$$x(t) = \sum_{l=1}^{L} A_q \cos(\omega_l t + \theta_l) \tag{3.9}$$

where $L$ is the maximum number of sines to be summed, and $\omega_l$ is the radian frequency of each sine. Actually, we can create a multi-sine by considering that $\omega_l$ has any frequency value, and thus we have a non-uniform multi-sine, which can be periodic or not. It will be periodic only if we manage to obtain a value of radial frequency $\Delta\omega$ that is the maximum common divider of each of the frequencies. In this case $\omega_l = k\ \Delta\omega$, where $k$ in this case is an integer. However, if, for instance, each of the sines is created with a different generator without any synchronization between them, the summation of $\omega_q$ will always create a non-periodic signal, since the phases will impose an almost periodic behavior.

Nevertheless, the sines can be chosen to be equally spaced as in $\omega_q = \omega_0 + (q-1)\Delta\omega$, with $\omega_0$ the position of the first tone and $\Delta\omega$ the constant frequency separation between them, and in that case the multi-sine is said to be uniformly spaced. The period of the multi-sine in this case will be the one expressed by

$$T_{\mathrm{MS}} = \frac{1}{\Delta\omega} \tag{3.10}$$

In order to extract some useful information from these multi-sine signals it is necessary to first investigate their capabilities.

First of all, it is interesting to know that those signals can present different time-domain waveforms, depending on the phase and amplitude selection for each tone. An illustration is presented in Fig. 3.14. This difference has driven microwave engineers to think about a way to characterize this type of time-domain evolution by defining an amplitude probability density function (pdf($x$)) to describe the signal behavior. Thus each of those signals will present different statistical values for each realization.

DEFINITION 3.1 *The probability density function,* pdf($x$), *is a function that describes the relative probability of a random variable existing at a given point in the space of observation. The probability of a random variable falling within a given set is given by the integral of its density over the set.*

This means that a careful design of the multi-sine to be used in the characterization of a nonlinear DUT is fundamental for a correct description of the system states.

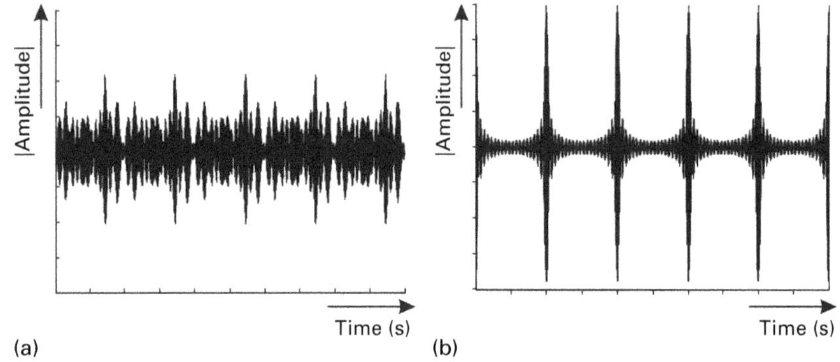


**Figure 3.14** Time-domain waveforms of two signals composed of ten evenly spaced tones of equal amplitude: (a) independent tones with a randomized phase arrangement and (b) all ten tones phase-locked to a common reference.

#### 3.4.1.1 Multi-sine with predetermined statistics

The response of nonlinear memoryless systems to the excitation's value is instantaneous. So in principle we could state that their instantaneous output is completely determined by the domain of stimulus amplitudes. However, most of the system's output metrics, such as the output power, power spectral density (PSD), and adjacent-channel power ratio (ACPR), have a statistical-average nature. Thus, just as important as the range of amplitude values covered by the output is the probability with which they are reached. This leads us to the intuitive thought that, on average, what matters is not the instantaneous amplitude itself, but the value weighted by pdf$(x)$.

For instance, although a certain very high instantaneous amplitude can determine the signal's amplitude span, it will actually become almost irrelevant to the system's output if the associated pdf$(x)$ is very low. Moreover, manufacturers of AWGs are starting to include in their equipment datasheets information on pdf$(x)$, or, more commonly, the complementary cumulative distribution function, ccdf$(x)$. Figure 3.15 presents some real ccdf$(x)$ curves for real wireless signals.

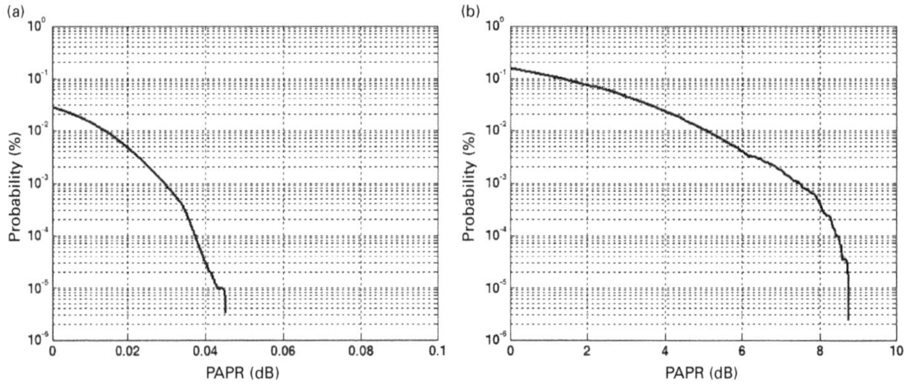


**Figure 3.15** The ccdf$(x)$ of real wireless communications signals, namely (a) GSM and (b) WiFi signals.

DEFINITION 3.2 *The* ccdf($x$) *is actually the complement of the cumulative distribution function,* cdf($x$)*, which is given by the integral of* pdf($x$)*, thus* ccdf($x$) $=$ $1 -$ cdf($x$).

To find the correct signal able to excite the different states and dynamics of the nonlinear DUT, several methods and algorithms can be used to design different multi-sines presenting different statistical behaviors.

### 3.4.1.2 Approximating the multi-sine pdf

Different multi-sines can present different signal statistics despite having the same power spectral density and average power. Figure 3.16 presents the pdf($x$) of two signals commonly used for model extraction and validation.

In Fig. 3.16, two multi-sines with uniform and Gaussian pdf($x$) are presented. These multi-sines have exactly the same power spectral density and average power, and their use is equivalent to testing the system with uniformly or Gaussian distributed band-limited white noisy waveforms of total power equal to the two-tone signal.

It is possible to design a specially tailored multi-sine to present a specific pdf($x$) pattern. In order to do that, an algorithm is used so that the multi-sine's pdf($x$) approximates the pdf($x$) of previously synthesized noise sequences [7, 8]. An outline of the algorithm is given below.

ALGORITHM 3.2 (multi-sine design with predetermined statistics)
*1. Synthesize a noisy signal with the specified* pdf($x$) *and re-order its instantaneous amplitude values in ascending order. This creates the vector of* pdf($x$) *bins for the noise.*
*2. Synthesize an equal-amplitude multi-sine with the prescribed number and frequency position of tones.*

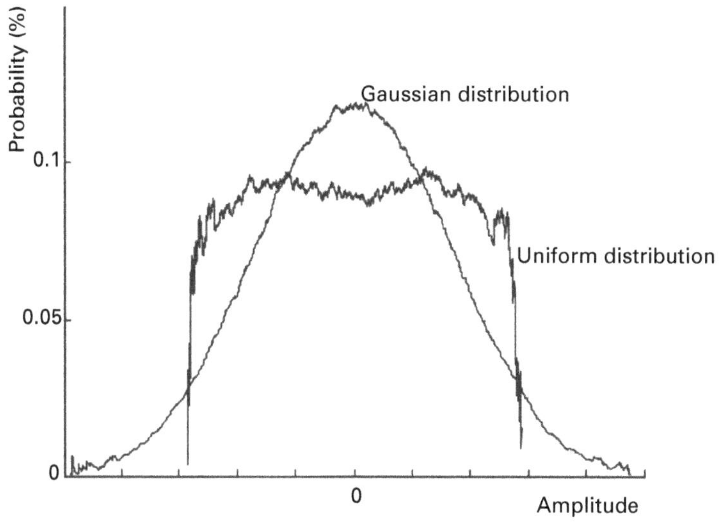


**Figure 3.16** The pdf($x$) of two multi-sines of uniform and Gaussian distribution, all with the same integrated power. © IEEE.

*3. Re-order its instantaneous amplitude values in ascending order, recording the time samples where they stood. This creates the vector of* pdf($x$) *bins for the multi-sine.*
*4. Substitute for amplitudes of the multi-sine vector of* pdf($x$) *bins that of the noise.*
*5. Restore these amplitudes in the original time samples of the multi-sine, creating a new multi-sine with the desired* pdf($x$).
*6. Calculate the DFT of this signal, and level off the resulting tone amplitudes, so that the total power is kept, maintaining the phases obtained. This is the desired multi-sine we sought.*
*7. If the process of tone-amplitude leveling off has modified the multi-sine* pdf($x$) *to an unacceptable error, repeat the application of the algorithm, using as the starting multi-sine the one that was synthesized in this way, until an acceptable error is reached.*

In Fig. 3.17(a) the original time behavior of the noisy signal can be seen. The noisy signal is generated with the selected target statistical behavior. Figure 3.17(b) presents the corresponding spectrum. Figure 3.17(c) presents the sorting of the signal amplitudes in ascending order. These figures correspond to point 1 of the algorithm.

Figure 3.17(d) is the corresponding time behavior of the generated uniform multi-sine with equal amplitude in each tone and 0° phase difference between the tones, corresponding to item 2 in the algorithm. Figure 3.17(e) is the corresponding spectrum, and Fig. 3.17(f) presents the sorting of the amplitudes as proposed in item 3 of the algorithm. Figure 3.17(g) is the spectrum after the amplitude change as proposed in items 4 and 5 of the algorithm. Finally, Fig. 3.17(h) is the result obtained for the multi-sine signal after the full algorithm has been implemented several times.

### 3.4.1.3 Multi-sine with predetermined higher-order statistics

For a memoryless nonlinearity first-order statistics like the pdf($x$) and its associated moments suffice for describing the integrated value of the distortion power.

DEFINITION 3.3 *The nth moment of the probability distribution can be defined as the value of the integral above the probability density function and can be given by* $u_n = \int_{-\infty}^{\infty} (x - c)^n f(x) dx$, *where c is the signal mean and* $f(x)$ *is its* pdf($x$).

Nevertheless, it can be proven [8] that this description does not suffice for signal characterization in nonlinear dynamic systems, since it is possible to generate multi-sines with similar pdf($x$) function description, but with a different output spectral regrowth mask. That is why an alternative multi-sine design technique is necessary for dynamic nonlinear systems.

Since in nonlinear dynamic systems the output does not change instantaneously with the input signal, the statistical relations of the output should include not only the static statistical behavior such as the pdf($x$) but also higher-order statistics [9]. If we consider the memory-polynomial description for the nonlinearity as in Section 1.4.1, the output can be described as a Volterra series:

$$y(t) = \sum_{n=1}^{N} y_n(t) \tag{3.11}$$

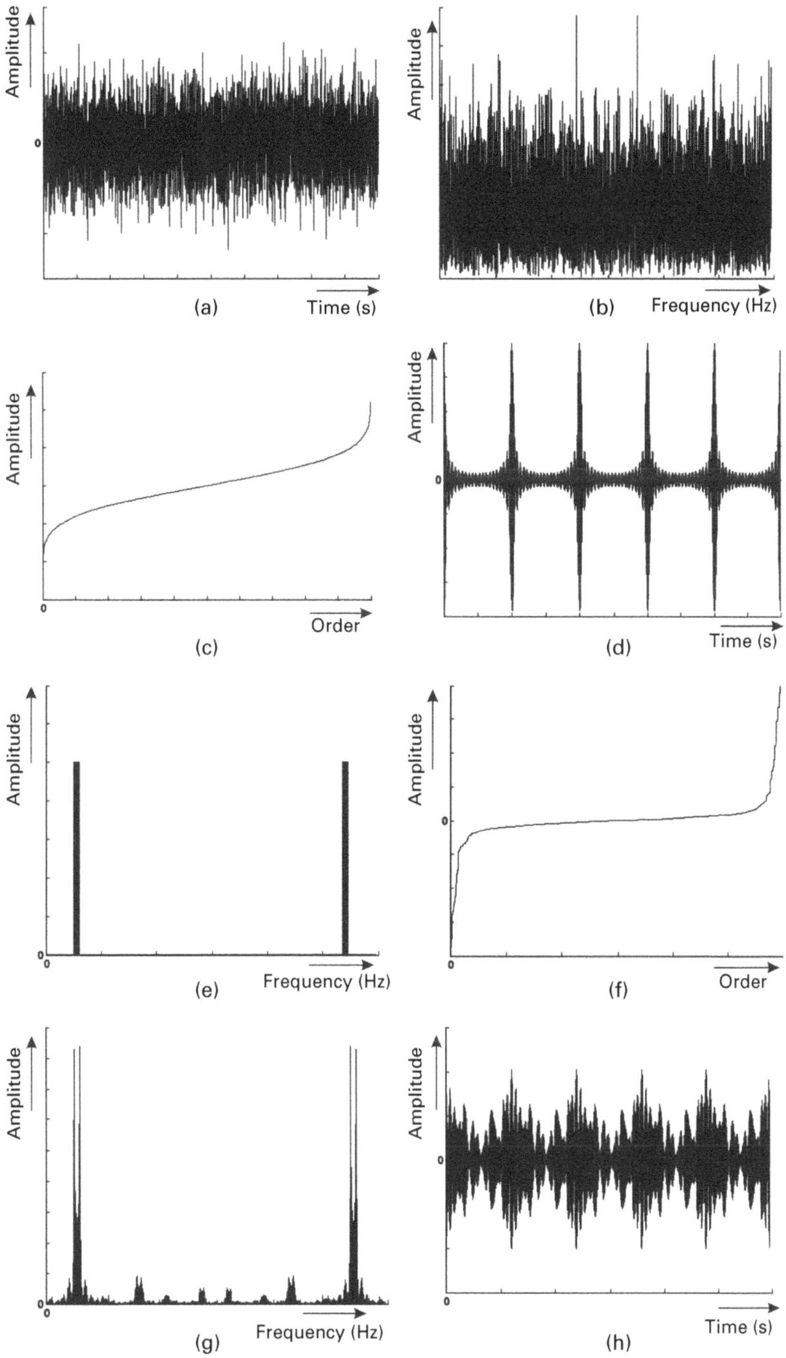


**Figure 3.17** Several steps relating to Algorithm 3.2. (a) Gausian noise signal time representation. (b) The spectrum of the Gaussian noise. (c) The amplitude ordering of the Gaussian signal. (d) A multi-sine with equal-amplitude time-domain behavior. (e) The spectrum of the multi-sine. (f) The amplitude ordering of the multi-sine. (g) The spectrum of the newly created multi-sine, after the change of amplitude and phase of the sine bins. (h) The final result corresponding to a multi-sine with a similar pdf to that of the original noisy signal. © IEEE.

where

$$y_n(t) = \int_{-\infty}^{+\infty} h_n(\tau_1, \ldots, \tau_n)x(t-\tau_1)\ldots x(t-\tau_n)d\tau_1 \ldots d\tau_n \tag{3.12}$$

and $h_n(\tau_1, \ldots, \tau_n)$ is the $n$th-order nonlinear operator.

Since the main idea is to obtain the output power spectral density (PSD) or the spectral mask of the output, $Y_{\mathrm{s},n}$ can be given by [9]

$$\begin{aligned} Y_{\mathrm{s},n}(\omega) &= \int_{-\infty}^{+\infty}\left\{\int_{-\infty}^{+\infty}\cdots\int_{-\infty}^{+\infty} h_n(\tau_1,\ldots,\tau_n)\left[\int_{-\infty}^{+\infty} X_{\mathrm{s}}(\omega_1)e^{j\omega_1(t-\tau_1)}\,d\omega_1\right]\right. \\ &\qquad \left.\cdots\left[\int_{-\infty}^{+\infty} X_{\mathrm{s}}(\omega_n)e^{j\omega_n(t-\tau_n)}\,d\omega_n\right]d\tau_1\ldots d\tau_n\right\}e^{-j\omega t}\,dt \\ &= \int_{-\infty}^{+\infty}\left\{\int_{-\infty}^{+\infty}\cdots\int_{-\infty}^{+\infty}\left[\int_{-\infty}^{+\infty}\cdots\int_{-\infty}^{+\infty} h_n(\tau_1,\ldots,\tau_n)\right.\right. \\ &\qquad \left.\times\, e^{j\omega_1\tau_1+\cdots+j\omega_n\tau_n}\,d\tau_1\,d\tau_n\right] \\ &\qquad \left.\times\left[X_{\mathrm{s}}(\omega_1)\ldots X_{\mathrm{s}}(\omega_n)e^{j(\omega_1+\cdots+\omega_n)t}\right]d\omega_1\ldots d\omega_n\right\}e^{-j\omega t}\,dt \\ &= \int_{-\infty}^{+\infty}\left[\int_{-\infty}^{+\infty}\cdots\int_{-\infty}^{+\infty} H_n(\omega_1,\ldots,\omega_n)X_{\mathrm{s}}(\omega_1)\ldots X_{\mathrm{s}}(\omega_n)\right. \\ &\qquad \left.\times\, e^{(\omega_1+\cdots+\omega_n-\omega)t}\,d\omega_1\ldots d\omega_n\right]dt \end{aligned} \tag{3.13}$$

The time integral is equal to unity if $\omega_1 + \cdots + \omega_n = \omega$ or if $\omega_n = \omega - (\omega_1 + \cdots + \omega_{n-1})$, otherwise it is zero. Using these simplifications, Eq. (3.13) becomes

$$\begin{aligned} Y_{\mathrm{s},n} &= \int_{-\infty}^{+\infty}\cdots\int_{-\infty}^{+\infty} H_n\big[\omega_1,\ldots,\omega_{n-1},\omega-(\omega_1+\cdots+\omega_{n-1})\big] \\ &\quad \times X_{\mathrm{s}}(\omega_1)\ldots X_{\mathrm{s}}(\omega-(\omega_1+\cdots+\omega_{n-1}))d\omega_1\ldots d\omega_{n-1} \end{aligned} \tag{3.14}$$

Using this expression we can calculate the output power density function, $S_{yy}(\omega)$, as

$$\begin{aligned} S_{yy}(\omega) &= \sum_{n_1=1}^{N}\sum_{n_2=1}^{N}\int_{-\infty}^{+\infty}\cdots\int_{-\infty}^{+\infty} H_{n_1}\big[\omega_1,\ldots,\omega-(\omega_1+\cdots+\omega_{n_1-1})\big] \\ &\quad \times H_{n_2}\big[\upsilon_1,\ldots,\upsilon-\big(\upsilon_1+\cdots+\upsilon_{n_2-1}\big)\big]^* \\ &\quad \times E\{X_{\mathrm{s}}(\omega_1)\ldots X_{\mathrm{s}}(\omega-(\omega_1+\cdots+\omega_{n_1-1}))X_{\mathrm{s}}(\upsilon_1)^* \\ &\qquad \ldots X_{\mathrm{s}}(\upsilon-(\upsilon_1+\cdots+\upsilon_{n_2-1}))^*\}d\omega_1\ldots d\omega_{n_1-1}\,d\upsilon_1\ldots d\upsilon_{n_2-1} \end{aligned} \tag{3.15}$$

For a third-order nonlinearity the output PSD will then be

$$
\begin{aligned}
S_{yy}(\omega) &= |H_1(\omega)|^2 S_{xx}(\omega) \\
&\quad +2\,\mathrm{Re}\left\{\int_{-\infty}^{+\infty}\int_{-\infty}^{+\infty} H_1(\omega)H_3(\omega_1,\omega_2,\omega-\omega_1-\omega_1)^* \right.\\
&\qquad \left. \times E[X(\omega)X(\omega_1)^*X(\omega_2)^*X(\omega-\omega_1-\omega_2)^*]d\omega_1\,d\omega_2\right\} \\
&\quad +\int_{-\infty}^{+\infty}\cdots\int_{-\infty}^{+\infty} H_3(\omega_1,\omega_2,\omega-\omega_1-\omega_1)H_3(\upsilon_1,\upsilon_2,\omega-\upsilon_1-\upsilon_1)^* \\
&\quad \times E[X(\omega_1)X(\omega_2)X(\omega-\omega_1-\omega_2)X(\upsilon_1)^*X(\upsilon_2)^* \\
&\quad \times X(\omega-\upsilon_1-\upsilon_2)^*]d\omega_1\,d\omega_2\,d\upsilon_1\,d\upsilon_2
\end{aligned}
\tag{3.16}
$$

In fact, since $H_n(\omega_i,\ldots,\omega_j)$, the nonlinear frequency operator, is well known because it depends exclusively on the DUT nonlinear model, only the value of the average $E\left[X(\omega_1)X(\omega_2)X(\omega-\omega_1-\omega_2)X(\upsilon_1)^*X(\upsilon_2)^*X(\omega-\upsilon_1-\upsilon_2)^*\right]$ should be equal for the multi-sine signal and the real signal that we want to describe. That is why the signals to approximate should obey the following formula, for a third-order nonlinearity:

$$
S_{xxxx}(\omega_1,\omega_2,\omega_3) = E\big[X(\omega_1)X(\omega_2)X(\omega_3)^*X(\omega_1+\omega_2-\omega_3)^*\big] \tag{3.17}
$$

Expressions like Eq. (3.17) are known as the signal's higher-order statistics [10] because they can be understood as being higher-order extensions of the first-order PSD, $S_{xx}(\omega)$.

We can thus conclude that two signals, $x(t)$ and $x_r(t)$, are similar up to order $n$, in the sense that they will present similar PSDs, if they have similar $n$th-order spectra $S_{xx}(\omega) \approx S_{x_r x_r}(\omega)$, $S_{xxxx}(\omega_1,\omega_2,\omega_3) \approx S_{x_r x_r x_r x_r}(\omega_1,\omega_2,\omega_3),\ldots, S_{x\ldots x}(\omega_1,\ldots,\omega_{n-1}) \approx S_{x_r\ldots x_r}(\omega_1,\ldots,\omega_{n-1})$.

The main problem in this multi-sine design is the high number of unknowns to be matched. For instance, if the noise is sampled as a $(2K+1)$-point DFT, $S_{xx}(\omega)$ involves only an average made over $2K+1$ complex numbers per signal realization. On the other hand, a second-order analysis $S_{xxxx}(\omega_1,\omega_2,\omega_3)$ already involves an average over $(2K+1)^3$ complex entities and, for the third order, $S_{xxxx}(\omega_1,\omega_2,\omega_3,\omega_4,\omega_5)$ must be estimated by averaging complex matrices of size $(2K+1)^5$. This implies that the design algorithm for this type of signal metric is extremely heavily computational. In fact, if on the one hand a lower number of sines is better for implementing the algorithm, on the other it restricts the number of unknowns and thus the number of free states. So a better approach is to consider a large number of sines, but to approximate the PSD only in some selected bins. In fact, this is exactly what most RF designers do, when they simulate/measure the DUT with a large number of pseudo-random samples, which corresponds to a large number of different multi-sines. Figure 3.18 presents this idea.

The design algorithm to generate this multi-sine is simply the optimization of the PSD of the multi-sine and the real signal in those bins by restoring to the higher-order

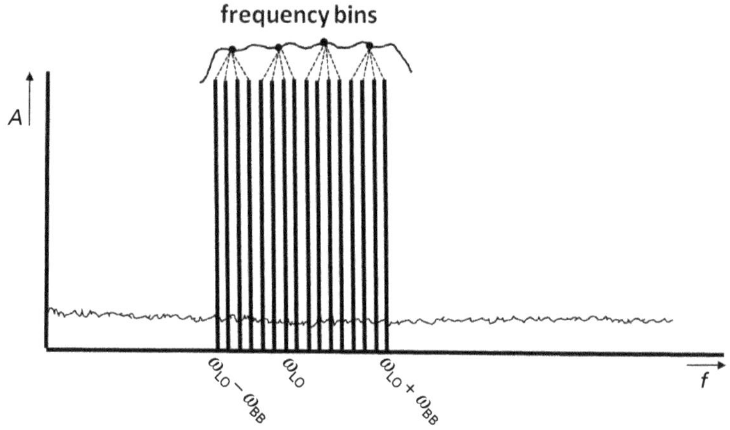


**Figure 3.18** The original signal's power spectral density function and the approximating multi-sine matched in a certain number of predefined frequency bins. © IEEE.

statistics functions. For instance, for a third-order nonlinearity, the errors to be minimized are expressed by

$$\varepsilon_{4,b_1,b_2,b_3} = \sum_{m_1=\frac{(b_1-1)M}{B}+1}^{\frac{(b_1)M}{B}} \sum_{m_2=\frac{(b_2-1)M}{B}+1}^{\frac{(b_2)M}{B}} \sum_{m_3=\frac{(b_3-1)M}{B}+1}^{\frac{(b_3)M}{B}} \{A_{m_1}A_{m_2}A_{m_3}e^{[\phi_{m_1}\phi_{m_2}-\phi_{m_3}-\phi_{m_1+m_2-m_3}]}\} \sum_{k_1=\frac{(b_1-1)M}{B}+1}^{\frac{(b_1)M}{B}} \sum_{k_2=\frac{(b_2-1)M}{B}+1}^{\frac{(b_2)M}{B}} \sum_{k_3=\frac{(b_3-1)M}{B}+1}^{\frac{(b_3)M}{B}} E[X(\omega_{k_1})X(\omega_{k_2})X(\omega_{k_3})^*X(\omega_{k_1}+\omega_{k_2}-\omega_{k_3})^*] \tag{3.18}$$

with $b_1, b_2, b_3 \in \{1, \ldots, B\}$.

Reference [11] gives more information on, and examples of, the design of this type of multi-sine.

### 3.4.1.4 Multi-sine generation mechanisms

Multi-sine generation can be done with several generators as in the two-tone case. The control of each sine phase is, however, very difficult or almost impossible, so this approach is used preferentially for generating a Gaussian-behaved multi-sine.

The most common approach involves the use of an AWG, similarly to what was presented for the two-tone case, whereby the signal is first created digitally and then up-converted to the RF output stage. In this case the first step to be undertaken consists of the generation of the intended multi-sine signal by first determining the amplitude and phase of each multi-sine bin via a computer software program, using the algorithms described previously.

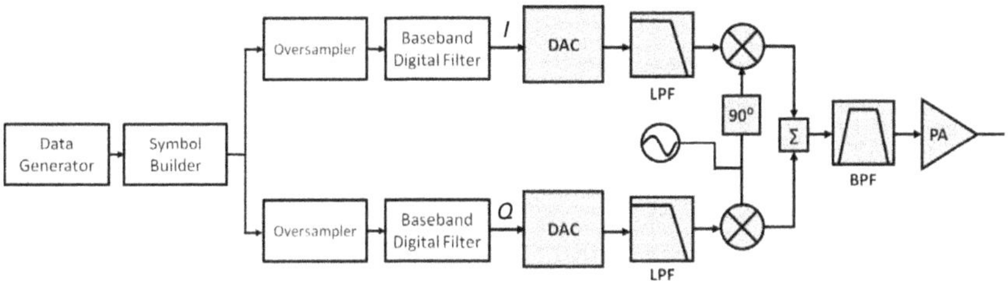


**Figure 3.19** The architecture of an arbitrary-waveform generator.

The output of these algorithms is the set of amplitudes and phases for each of the multi-sine bins, resulting in the digital version of the real ($I$) and imaginary ($Q$) components of a periodic low-pass baseband signal sampled at a predetermined frequency. This digital signal is then downloaded into the physical memory of an RF AWG, as presented in Fig. 3.19. The approach is similar to the one presented in Fig. 3.12, but in this case the signal is actually created previously in a processor and then fed to an $I/Q$ modulator.

As depicted in the AWG block diagram in Fig. 3.19, this digital $I/Q$ baseband signal is then converted to the analog domain via two independent DACs, and then upconverted onto the carrier frequency in the $I/Q$ modulator. Finally, the bandpass RF signal is amplified and presented at the output of the AWG.

The multi-sine signal is now generated and the same problems as arose before in Section 3.3.1 apply also to this case. If the signal is wideband, the non-idealities of the AWG can significantly degrade the intended multi-sine at the output of the generator.

Actually, the phase arrangement that was previously designed in the computer software may be completely changed, thereby degrading the overall measurement of the DUT. In order to correct for this problem some authors [12] presented a corrected multi-sine waveform generator that uses a feedback loop to synthesize the correct waveform at the input of the DUT.

The implemented correction algorithm works by sampling the output of the generator and comparing the amplitudes and phases of the tones in the real multi-sine with the proposed amplitudes and phases designed in the computer software. In order to do that, a time-domain scope could be used, by applying an FFT afterwards to the obtained result. Unfortunately, the phase determination is particularly difficult to obtain, because the instant of acquisition is random (asynchronous triggering), and therefore the raw values of the phases of the desired spectral components are also random. However, if a time alignment is performed on each record, the phase differences between the tones become deterministic, enabling the sought comparison between the components of the measured signal and the phases stored for the desired signal. In the optimum case, in which a perfect compensation is achieved, these differences should be zero.

Section 4.2.3.4 presents some of the techniques used for phase alignment. The proposed multi-sine downloaded to the AWG is then changed such that the multi-sine at the output matches the one that was targeted. Figure 3.20 presents the block diagram used

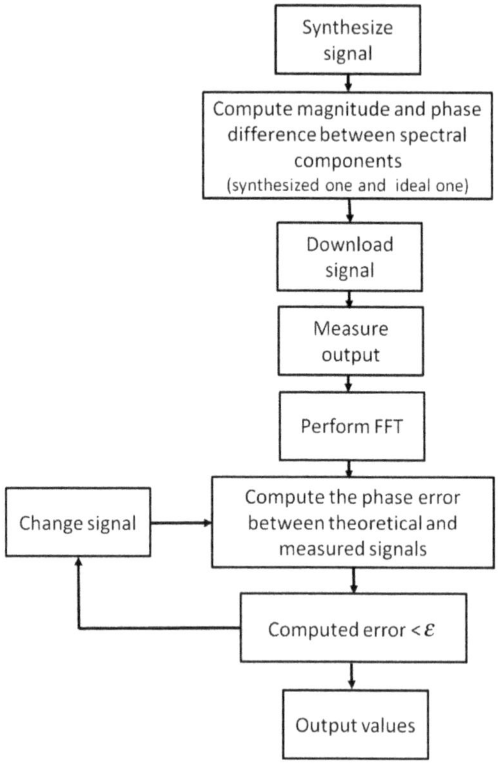


**Figure 3.20** The multi-sine correction algorithm.

for this matching measuring scheme. Figure 3.21 shows the corresponding experimental setup.

Moreover, and despite the fact that the description of the operation has concentrated on the signal's fundamental spectral components, the suggested algorithm and setup can also minimize any spectral regrowth arising in the instrument's RF front end and in the generator-to-DUT signal path, since the proposed control is actually implementing some form of adaptive digital pre-distortion linearization [13].

### 3.4.2 Complex modulated signals

Modulated signals are in fact the core type of characterization excitation that most of the new wireless systems use, since they are closer to the real environments in which most RF systems operate. Nevertheless, from an RF point of view, the generation of this type of excitation is similar to what was previously used for multi-sine signals, as presented in Fig. 3.19. The implementation of each modulated signal format is nothing more than a baseband algorithm that should fill up the look-up table according to a certain type of codification, which is normally phase-quadrature modulated.

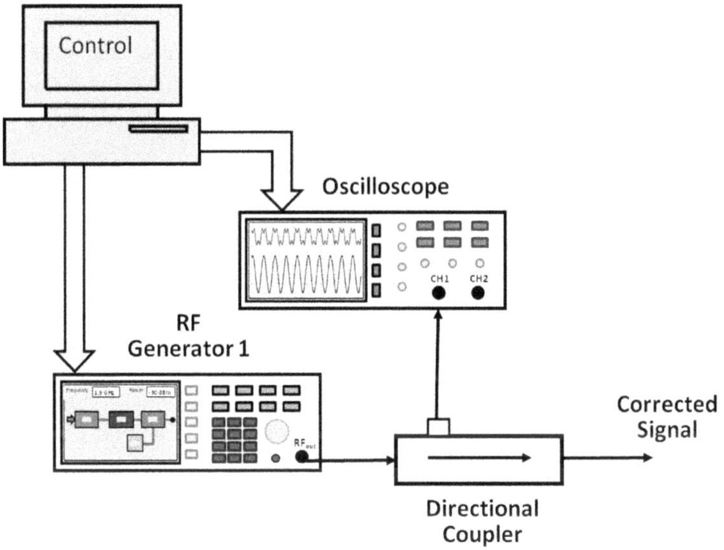


**Figure 3.21** The experimental setup for the corrected "ideal-waveform generator."

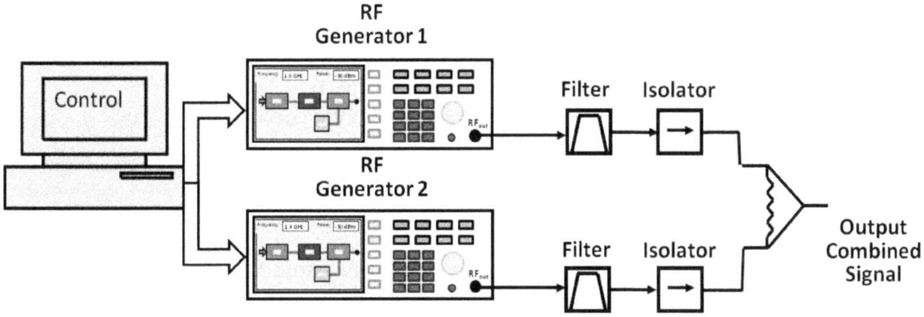


**Figure 3.22** The generation of multiple modulated signals.

The user should understand the bandwidth and the peak power that the generated modulated signal is going to create, since that will impose a strong restriction on the RF components that will be used afterwards, and of course it will also impose several drawbacks on the DACs to be used.

Moreover, a matter of importance in RF generation is the possibility of using several modulated signals combined in order to mimic the RF interference that can be found in many practical wireless systems. This type of signal can be generated using a special algorithm at baseband and then using a scheme similar to the ones presented in Fig. 3.19 to upload it, but in that case the maximum bandwidth of the signal to be generated is on the order of the limitations of the RF generator. Therefore, another approach involves combining several AWGs, as illustrated in Fig. 3.22.

In this case the signals are actually generated individually by two AWGs and up-converted to different RF carriers, namely the filter and isolator in each path as the same function, as was explained in Section 3.3.1 for the two-tone case. It should also be pointed out that the combination of the two signals will create a higher PAPR, and that in a real situation the carriers should be completely uncorrelated; that is, they should not have a common clock. Nevertheless, if that is needed, then an RF reference should be used to synchronize the two carrier clocks. This approach is actually extensively used in MIMO characterization systems.

## 3.5 Chirp signals

Another signal that is gaining importance in RF laboratories is the chirp signal. This is a signal whose frequency varies through time. Chirp signals allow the mimicking of switched-mode real signals, such as, for instance, any time-division-duplex (TDD) system.

One of the examples is the two-tone chirp signal that allows one to gather information about the nonlinear distortion and the nonlinear dynamic distortion of devices [14]. This signal waveform can be mathematically described by

$$x(t) = A\cos\left[2\pi\left(f_c - \frac{\delta(t)}{2}\right)t\right] + A\cos\left[2\pi\left(f_c + \frac{\delta(t)}{2}\right)t\right] \quad (3.19)$$

As can be seen, the tone spacing in this function varies with time $t$.

One of the possible implementations of that function could be

$$\delta(t) = \sum_{k=0}^{N} \rho_k \operatorname{rect}(t - t_k) \quad (3.20)$$

The value of $t_k$ should be the minimum possible for a correct evaluation of the signal spectrum in the timed waveform. In fact the evaluation will be made possible only by using a time–frequency transform such as the short-time–frequency transform presented in Section 2.5 and in [14]. The plots in Fig. 3.23 present the RF time and spectrum domains of the chirp waveform for the overall sampled time.

To construct this waveform, an AWG can be used, decomposing the waveforms into a complex envelope representation,

$$x(t) = |\tilde{z}(t)|\cos[\omega_c(t) + \angle\tilde{z}(t)] \quad (3.21)$$

where

$$\tilde{z}(t) = e^{-j2\pi\frac{\delta(t)}{2}} + e^{j2\pi\frac{\delta(t)}{2}} \quad (3.22)$$

The baseband interpretation of this modulation is that of two single-sideband tones moving away from each other over time as illustrated in Fig. 3.24. The idea is that the two tones and the intermodulation products can be measured and characterized over wide tone spacings with a single measurement using a VSA. Section 4.2.6.5 presents more information about this type of bench.

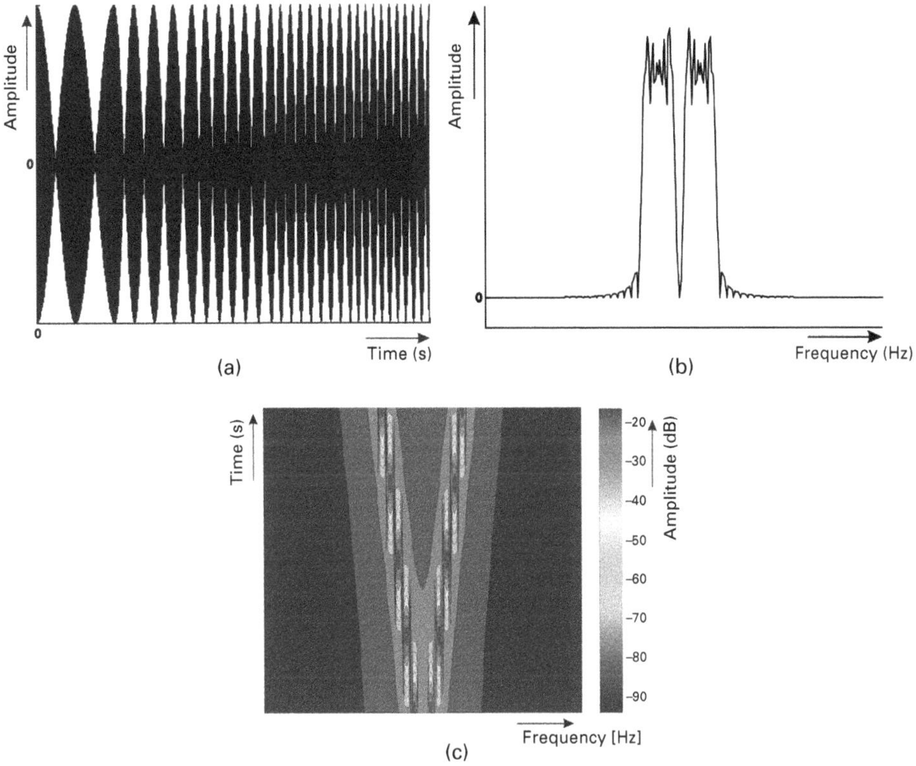


**Figure 3.23** (a) The time domain, (b) the spectrum, and (c) the real-time spectrum of the proposed varying two-tone waveform.

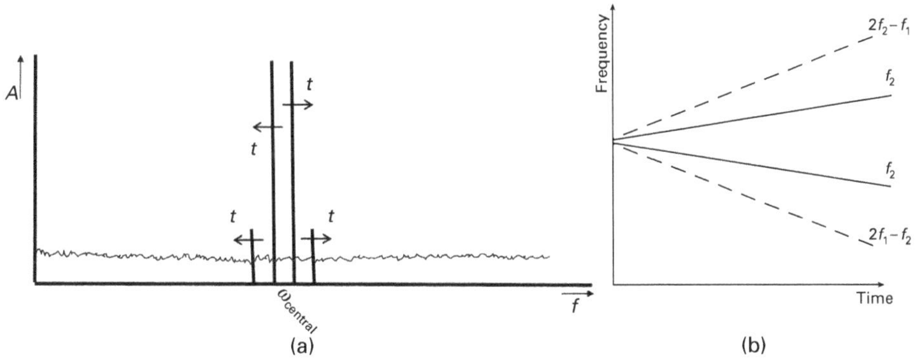


**Figure 3.24** A graphical explanation of the dynamic frequency two-tone waveform generation.

## 3.6 Comb generators

As noted previously in Section 2.7, an NVNA strongly depends on a comb generator. The comb generator was developed to provide precise phase calibration for an NVNA's instruments. A comb generator produces nothing more than a sinusoidal wave and its

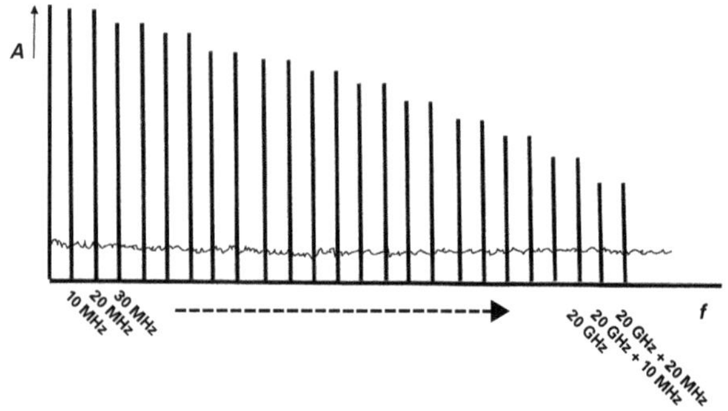


**Figure 3.25** The output spectrum components of a comb generator.

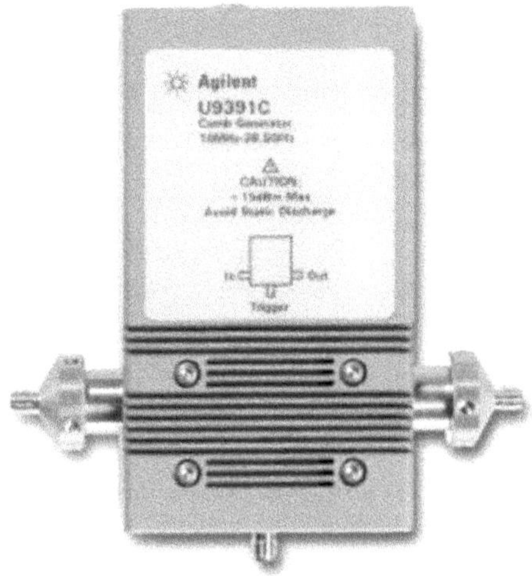


**Figure 3.26** A commercial comb generator made by Agilent Technologies. © Agilent Technologies.

harmonics, the objective being that the output contains a huge number of harmonics from low to high frequency. Most comb generators allow the generation of tones of frequency 10 MHz to 20 or 50 GHz with 10-MHz spacing. Figure 3.25 presents the typical spectrum content of a comb generator.

The main principle involved to building a comb generator is creating a pulse response that should be so narrow in the time domain that it implies a huge bandwidth in the frequency domain. The repetition period of that pulse will impose the frequency spacing on the spectrum content. A step recovery diode is most commonly used for designing this type of comb generator [15]. Figure 3.26 presents a typical commercial comb generator.

## 3.7 Pulse generators

The final type of generator that is covered in this chapter is pulse generators. Pulse generators have a wide range of uses in connection to wireless applications. They can be employed from device level up to system level.

The use of pulse generators at system level is well established for applications such as pulsed radar, but it is more recent in the field of wireless applications. The typical application nowadays is ultra-wideband (UWB) communication to enable higher data rates. The specifications of such UWB pulse generators are different from those of the pulse generators used for device characterization, which will be discussed next, because the spectral mask is strictly regulated by standardization bodies, such as by the standard IEEE 802.15.

At device level, pulsed excitation is adopted to evaluate semiconductor technologies for the presence of memory effects, such as traps and thermal heating. The most common uses are pulsed DC measurements and pulsed $S$-parameter measurements. The principle is that the duration of the pulses is sufficiently short that memory effects are not excited. In other words, if there is a difference between regular DC and pulsed DC measurements, and similarly as regarding $S$-parameter measurements, then memory effects are present. Memory effects are to be avoided in wireless applications since they complicate design at circuit and system level. For example, the linearization of a power amplifier becomes troublesome if the amplifier manifests memory effects. When selecting pulse generators for device characterization, the need for an external bias tee should be avoided, since a bias tee and the related additional cabling change the shape of the pulse, which may render interpretation of the measured data difficult. Some examples of pulse generators are depicted in Fig. 3.27. If they are to be used for device characterization, the following specifications are of importance.

- Amplitude of the pulse. This should be adjustable to cover the full operating range of the device. It is typically specified for a 50-Ω load.
- DC offset. This should be adjustable, and both negative and positive values should be possible.
- Pulse width. This should be adjustable, such that various memory effects can be examined; a typical range is 20 ns to 1 ms.
- Duty cycle. This relates to the time between pulses; it is the ratio of the pulse duration to the duration of the total period, before the next cycle starts with the next pulse. It should be very small (e.g., 0.01% or even less), so that the device returns to steady-state condition before the next pulse arrives.
- Rise time of leading edge. This is usually expressed as the time required to go from 10% of the value to 90%. It should be on the order of 1 ns or less.
- Fall time of trailing edge. This is the time required to go from 90% of the value to 10%. It should be on the order of 1 ns or less.
- Pulse repetition frequency. This typically ranges from less than 1 Hz to 1 MHz.

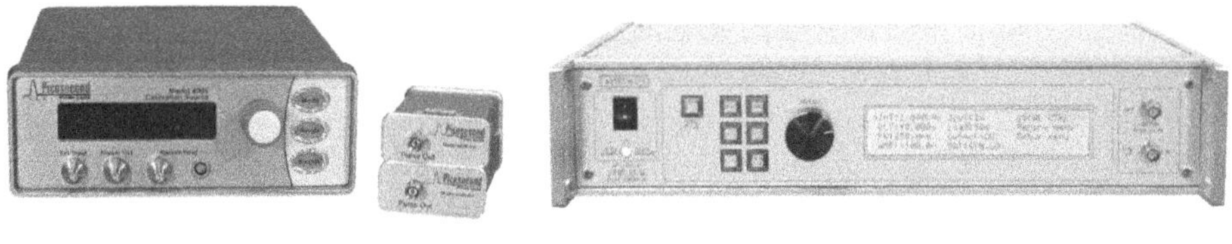

**Figure 3.27** Examples of pulse generators. © vendors.

Pulse generators can be designed in various ways. The main principle is to create a strongly nonlinear device, because a rich harmonic content in the spectral domain corresponds to a narrow pulse in the time domain. Traditional approaches make use of step-recovery diodes (SRDs) and nonlinear transmission lines [16], but over the years other designs have been developed by researchers and manufacturers, such as exploiting the step-recovery effect in bipolar transistors, adopting logic-gate switching, etc. In fact, the comb generator discussed in Section 3.6 is also a type of pulse generator.

## Problems

**3.1** Explain the main drawback of a high value of phase noise in the signal generator. What can be done to minimize it?

**3.2** Considering that we want to generate a two-tone signal with an arbitrary waveform generator with a frequency spacing of 10 kHz, please write down the algorithm that is needed in order to create such a waveform.

**3.3** Explain the differences in terms of signal generation between a non-uniform multi-sine and a uniformly spaced multi-sine, stressing when to use each one.

**3.4** What is the main difference between the generation of an envelope multi-sine with center frequency 0 Hz and the generation of one with a low-IF center frequency?

**3.5** If the output of my generator has an $IP_3$ of 50 dBm, calculate the maximum output power that can be generated for a minimum intermodulation ratio of IMR $>$ 40 dB.

**3.6** Implement a laboratory setup to generate a multi-sine signal with uniform statistical behavior.

**3.7** If a modulated signal has a PAPR of 10 dB, and we have an AWG with a maximum output power of 20 dBm, what is your first guess power to use?

**3.8** How can you measure a chirp signal using the instrumentation presented in Chapter 2.

**3.9** What are the main limitations on using an AWG for two-tone signal generation?

**3.10** What are the main drawbacks of using a multi-sine generator?

## References

[1] G. D. Vendelin, A. M. Pavio, and U. L. Rohde, *Microwave Circuit Design Using Linear and Nonlinear Techniques*. New York: Wiley, 2007.

[2] D. M. Pozar, *Microwave Engineering*. New York: Wiley, 2005.

[3] J. Klapper and J. Frankle, *Phase-Locked and Frequency Feedback Systems*. New York: Academic, 1972.

[4] J. C. Pedro and N. B. Carvalho, *Intermodulation Distortion in Microwave and Wireless Circuits*. New York: Artech House, 2003.

[5] M. B. Steer, *Microwave and RF Design: A Systems Approach*. Herndon, VA: SciTech Publishing, 2010.

[6] R. Pintelon and J. Schoukens, *System Identification: A Frequency Domain Approach*. New York: IEEE Press, 2001.

[7] J. Schoukens and R. Pintelon, *System Identification – A Frequency Domain Approach*. New York: Wiley 2001.

[8] J. Schoukens and T. Dobrowiecki, "Design of broadband excitation signals with a user imposed spectrum and amplitude distribution," in *IEEE Instrumentation and Measurement Technology Conference*, San Paul, USA, May 1998, pp. 1002–1005.

[9] J. C. Pedro and N. B. Carvalho, "On the use of multi-tone techniques for assessing RF components' intermodulation distortion," *IEEE Trans. Microwave Theory Tech.*, vol. 47, no. 12, pp. 2393–2402, Dec. 1999.

[10] J. C. Pedro and N. B. Carvalho, "Designing band-pass multisine excitations for microwave behavioral model identification," in *IEEE International Microwave Theory and Technology Symposium Digest*, Fort Worth, USA, Jun. 2004, pp. 791–794.

[11] J. C. Pedro and N. B. Carvalho, "Designing multisine excitations for nonlinear model testing," *IEEE Trans. Microwave Theory Tech.*, vol. 53, no. 1, pp. 45–54, Jan. 2005.

[12] N.B. Carvalho, J.C. Pedro, and J.P. Martins "A corrected microwave multisine waveform generator," *IEEE Trans. Microwave Theory Tech.*, vol. 54, no. 6, pp. 2659–2664, Jun. 2006.

[13] N. Pothecary, *Feedforward Linear Power Amplifiers*. Norwood, MA: Artech House, 1999.

[14] N. B. Carvalho, Jie Hu, K. G. Gard and M. B. Steer, "Dynamic time-frequency waveforms for VSA characterization of PA long-term memory effects," in *ARFTG Microve Measurements Conference*, Honolulu, Hawaii, Jun. 2007.

[15] Agilent Technologies, Agilent U9391C/F/G, Comb Generators, Technical Overview. 2011.

[16] M. Rodwell, M. Kamegawa, Y. Ruai *et al.*, "GaAs nonlinear transmission lines for picosecond pulse generation and millimetre-wave sampling," *IEEE Trans. Microwave Theory Tech.*, vol. 39, no. 7, pp. 1194–1204, 1991.

# 4 Test benches for wireless system characterization and modeling

## 4.1 Introduction

In this chapter the main objective is to present several test benches for modeling and characterization that allow a correct identification of several linear and nonlinear parameters useful for wireless systems. These benches include the following.

1. Test benches for characterization
   a. Power-meter measurement
   b. Noise-figure measurements
   c. Two-tone measurements
   d. VNA measurements
   e. NVNA measurements
   f. Modulated signal measurements
   g. Mixed-domain (analog and digital) measurements
   h. Temperature-dependent measurements
2. Test benches for behavioral modeling
   a. Volterra-series modeling
   b. State-space modeling
   c. Beyond *S*-parameters

## 4.2 Test benches for characterization

### 4.2.1 Power-meter measurements

As referred to in Chapters 1 and 2, power was the first important measurement to become available for radio communications, since it allows a wireless system engineer to calculate and predict some of the most important aspects of radio propagation. In order to measure power, a power meter should be used. A typical setup is presented in Fig. 4.1. A typical instrument is shown in Fig. 4.2.

When using a power meter, several steps should be executed prior to the measurement itself, and several parameters should be known in advance for tuning and selecting the correct set. The first step towards a power measurement is the correct selection of the power probe. The selection depends on the frequency range which the probe can cover, as well as the power range. In this first step the operator should also decide on whether

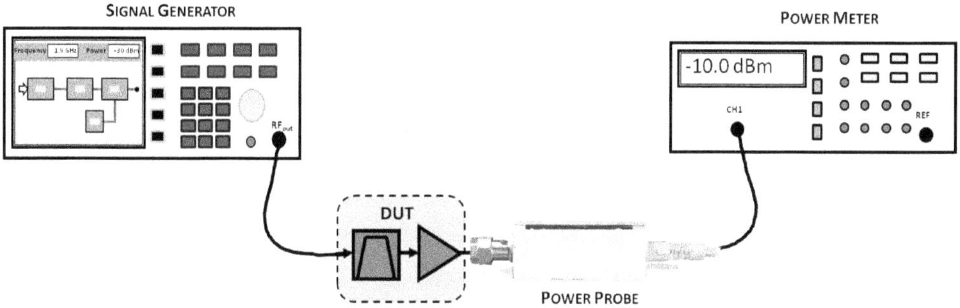


**Figure 4.1** A typical power-meter bench, presenting a signal generator, the DUT, a power probe, and the power-meter display itself.

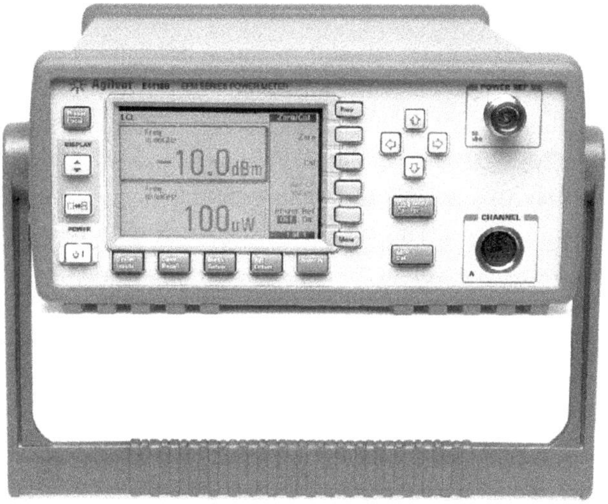


**Figure 4.2** A typical power meter, showing the calibration terminal and the probe terminal. © Agilent Technologies.

an average probe or a fast probe should be selected, for measuring the average power or the instantaneous envelope power, respectively.

The next step is related to the calibration of the power probe. It includes handling the calibration factors (see Section 2.2.6), which can be preloaded on the instrument or uploaded on an as-needed basis. This step allows the instrument to have knowledge of the correct calibration coefficients for that specific probe, and thus to eliminate any temperature and frequency mismatch arising from the probe. The next step is called zeroing, by which the instrument calculates the zero value of the power at its input. Be sure that at this time all the RF sources in the bench are switched off. During operation, one should carry out zeroing of the RF power probe several times during the day to compensate for any drifts, e.g., due to a slight change in environmental temperature. Do not forget that there should be no signal at the input when one is zeroing.

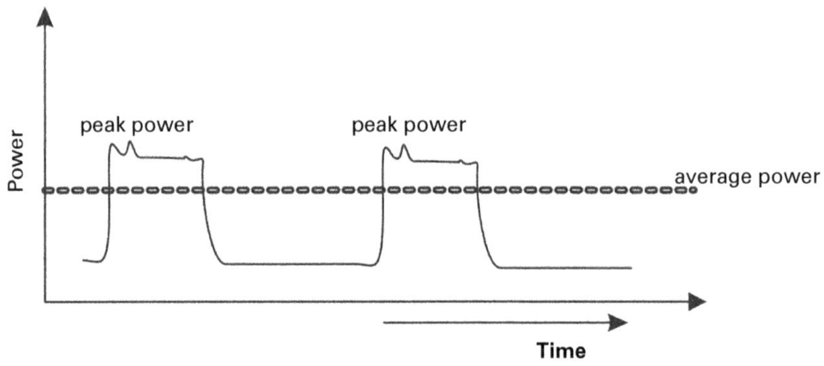


**Figure 4.3** The variation of peak power over time.

The subsequent step normally involves the calibration generator, which, as explained in Section 2.2.6, is nothing more than a known RF generator that is used to fine-tune the calibration coefficients. In most power meters, this implies that a very stable signal (normally of 50 MHz, 1.000 mW) is applied to the power probe as a reference level. The meter itself will adjust its gain to match the response of the power sensor.

After this step, the instrument is calibrated and a real power measurement can actually begin.

The procedure can be summarized as follows.

1. Warm up the measurement instrument for at least one hour prior to any measurement.
2. Reset all prior saved data.
3. Select the type of units that you want for your measurement, dBm or watts.
4. Put all your bench instrumentation in the RF off state, so that no RF power is being generated, and select the equipment zeroing stage.
5. Select the calibration factor for your power probe (in certain instruments this step is done prior to the calibration itself).
6. Calibrate the instrument using the internal RF reference generator.
7. Now disconnect all your RF power and 0 W should be measured.
8. The equipment can now start its measurement.

It should also be stated here that sometimes peak power is a very important problem to deal with. In a power measurement for a CW signal, the power remains constant over time, which means that the measured power maintains its value independently of the time window.[1] Nevertheless in new and emerging wireless signals, power varies with time, and sometimes we can identify peaks in the signal waveform. These peaks can degrade the operation of our system and severely hinder wireless communication. Figure 4.3 shows a typical wireless signal presenting a peak power.

[1] Considering that at least one period of the CW signal is acquired.

For measuring the peak power we should define correctly the time window within which we want to measure the power, and guarantee that our power probe has good enough dynamics to capture that power within the time window. We should stress here that power is actually an average value by its very nature, so for measuring the peak power the correct time window should be selected and the probe should respond rapidly within that time window:

$$P_s = \frac{1}{T}\int_{-T/2}^{T/2} |x(t)|^2\, dt \tag{4.1}$$

Sometimes when a fast power probe is not available an oscilloscope can also be used, but we should guarantee that it is capable of measuring the signal at the frequencies of interest. One important point that should be mentioned here is that a power meter actually measures all the power at the probe, which means that the power that is visible in the power-meter display includes the power across the full bandwidth of the probe, including any spurious signal coming out the measured DUT.

### 4.2.2 Noise-figure measurements

As mentioned in Section 2.10, the noise figure represents the noise added by a specific DUT, and, as outlined in that section, we have two possible approaches for implementing these measurements, namely including a noise source or not doing so.

The approach that does not include a noise source is implemented in the instrument itself using a specific mathematical algorithm as presented in Section 2.10.2, so its operational principle does not impose a specific measurement bench, but rather a specific instrument. In this section we will explain mainly the operation of the traditional noise-figure meter using a noise source.

Figure 4.4 presents a typical bench for measuring noise figures. The bench includes a noise source and a noise-figure meter (which is nothing more than a power meter that can be included in a spectrum analyzer).

The method identified in Fig. 4.4 is normally known as the $Y$-factor method, and, as specified in Section 2.10.1, it is also known as the hot and cold method, since it measures

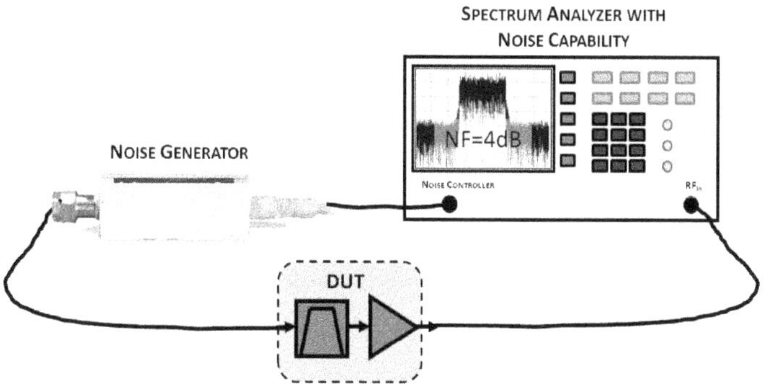


**Figure 4.4** The noise-figure measurement bench, where a spectrum analyzer, a noise generator, and the DUT are visible.

the DUT with a noise source connected to it (hot) and without the noise source (cold). So the procedure follows the following steps.

1. Connect the noise source directly to the measurement instrument.
2. Measure the hot and cold noise power, that is, the power with the noise generator switched on and off.
3. Save these values, since they will be used for the final measurements.
4. Connect the noise source to the input of the DUT.
5. Measure the output noise power of the DUT with the noise source switched on.
6. Measure the noise power of the DUT output again, with the noise source switched off.

The $Y_{\text{factor}}$, as in Section 2.10, will thus be

$$Y_{\text{factor}} = \frac{P_{\text{nDUTON}}}{P_{\text{nDUTOFF}}} = \frac{F_{\text{noisesource}} + F_{\text{overall}}}{F_{\text{overall}}} \tag{4.2}$$

As mentioned in Section 2.10.2, $F_{\text{overall}}$ is actually the combined noise factor of the DUT and the SA. The latter can be removed using the noise-figure Friis formula:

$$F_{\text{overall}} = \frac{F_{\text{noisesource}}}{Y_{\text{factor}} - 1} = F_{\text{DUT}} + \frac{F_{\text{SA}} - 1}{G_{\text{DUT}}}$$

whence

$$F_{\text{DUT}} = F_{\text{overall}} - \frac{F_{\text{SA}} - 1}{G_{\text{DUT}}} \tag{4.3}$$

where $F_{\text{DUT}}$ is the noise factor of the DUT, $F_{\text{noisesource}}$ is the noise factor of the noise source, $F_{\text{overall}}$ is the overall measured noise factor and $F_{\text{SA}}$ is the noise factor of the spectrum analyzer. $G_{\text{DUT}}$ can be obtained as explained in Section 2.10.

The noise figure, $\text{NF}_{\text{DUT}}$, can be calculated as the logarithmic equivalent of the noise factor.

#### 4.2.2.1 Noise-figure calibration

As was mentioned previously, for the calibration procedure the measurement instrument (SA or power meter) is fed with a noisy signal generated by the noise source, and the instrument measures the noise power when the noise source is on and off. At this time the instrument calculates the on power and off power, and resets the noise figure to 0 dB, waiting for the DUT to be measured. The aim of this calibration is mainly to reduce the internal noise figure of the instrument, but, if the instrument presents a front-end gain higher than 30 dB, the calibration procedure can be minimized.

The operator should also be aware that any extra noise sources around the instrument, for instance mobile phones, WiFi interfaces, fluorescent lights, etc., can completely degrade the measurement. So we should avoid laboratory spaces where such extra noise sources are present.

Moreover, any mismatch could also degrade your measurement. Any mismatch will degrade the calibration and thus the measured $Y$-factor. As presented in other schemes, the use of an isolator can help to minimize the uncertainty problem due to mismatches.

The same problems as identified with power meters can actually be considered here. For instance, if the measurement is overloaded, nonlinear behaviors can appear and may completely degrade the measured result. So the user should ensure that the input of the power meter (noise-figure analyzer) is not overloading the RF front end.

### 4.2.3 Two-tone measurements

As stated in Section 1.5.2, two-tone measurements are actually one of the most well-known types of test for the measurement of nonlinear distortion in RF and wireless circuits and systems. As the name states, the approach involves generating a two-tone signal that can be used as the excitation pattern. This two-tone generation mechanism could be realized using an AWG, or using two separate generators. In both cases the main idea is to have two sinusoidal signals without any type of harmonic behavior, Fig. 4.5.

#### 4.2.3.1 Two-tone signal generation using an arbitrary-waveform generator

As can be seen from Fig. 4.5(a), the spectra should be as clear as two Dirac delta functions in the frequency domain, and in the time domain we should have a clear envelope shape that is nothing more than a sinusoid at the beating frequency of the two tones, as illustrated in Fig. 4.5(b).

As can be observed and understood from Section 3.3, there are two ways to generate a two-tone signal. One method makes use of an AWG, in which case a baseband equivalent waveform is generated. This can be represented by

$$x_{\mathrm{BB}_{2\mathrm{tone}}} = A_1 e^{-j(\delta\omega+\theta_1)} + A_2 e^{+j(\delta\omega+\theta_2)} \tag{4.4}$$

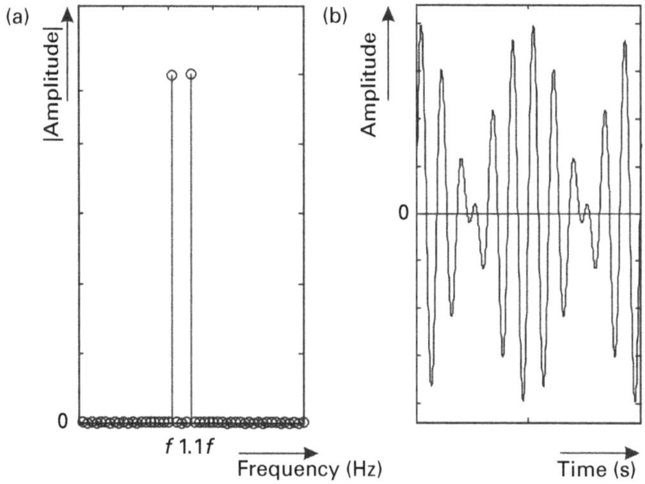


**Figure 4.5** Two-tone signal representation in the frequency and time domains: (a) two-tone spectra and (b) two-tone variation over time.

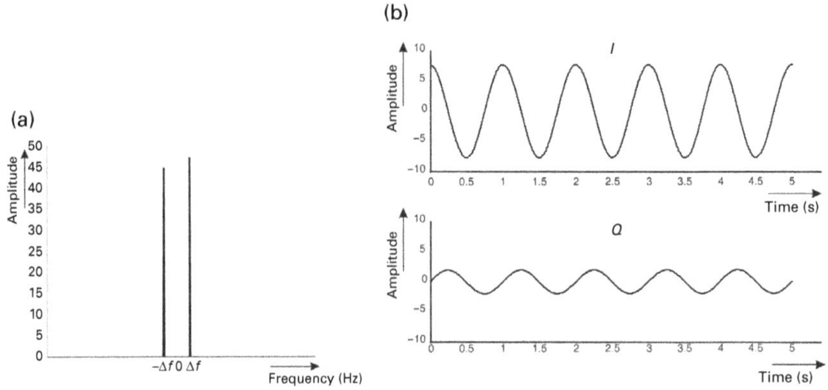


**Figure 4.6** (a) Two-tone signal generation using an AWG and (b) two-tone baseband complex behavior.

This envelope complex signal can then be up-converted to RF, and will be transformed into

$$x_{2\text{tone}} = \Re\left\{A_1 e^{-j(\omega_1+\theta_1)} + A_2 e^{-j(\omega_2+\theta_2)}\right\} \tag{4.5}$$

In the above equations, $\omega_1$ and $\omega_2$ are the tone frequencies, $\theta_1$ and $\theta_2$ represent the phases of the tones, $A_1$ and $A_2$ are the amplitudes of the tone, and $\delta\omega = (\omega_2 - \omega_1)/2$ is half of the tone-spacing frequency, and thus actually the frequency of the complex envelope is this beating frequency, as illustrated in Fig. 4.6(a). It should be stressed that the phase difference between the two tones is not significant for the signal shape of the time-domain waveform, in contrast to the multi-sine case as presented in Chapter 3.

Unfortunately, and as mentioned in Section 3.3.1, this generation mechanism implies that a large amount of spurious signal appear at the output, with the strongest signal being the local oscillator, which normally appears in between the two tones. So special care should be exercised when operating with a system like this. Despite that, the two-tone RF signal will also traverse the RF front end of the generator, so some nonlinear distortion leakage will also appear at the input of the DUT. Hence again the operator should take care not to corrupt the nonlinear-distortion measurement.

One possible arrangement in order to obviate this inherent nonlinear distortion is to create the signal at a lower amplitude, backing off the output generator amplifier, and then amplify the signal output with a better-quality external amplifier, one with a higher $IP_3$ value. This can obviate the generation of extra nonlinear distortion, but it unfortunately also generates an extra noise floor due to the amplification of the generator noise in the output external power amplifier. Figure 4.7(b) presents this configuration.

### 4.2.3.2 Two-tone signal generation using two CW generators

The second method employed to generate a two-tone signal is to use two CW generators, as sketched in Chapter 3. In that case we will combine one CW generator at a specific frequency with another one at another frequency. If both generators are locked to a common reference, then the phase difference between them can be changed. If the generators are not locked, the tones will be uncorrelated, and thus no phase relationship can be considered between them.

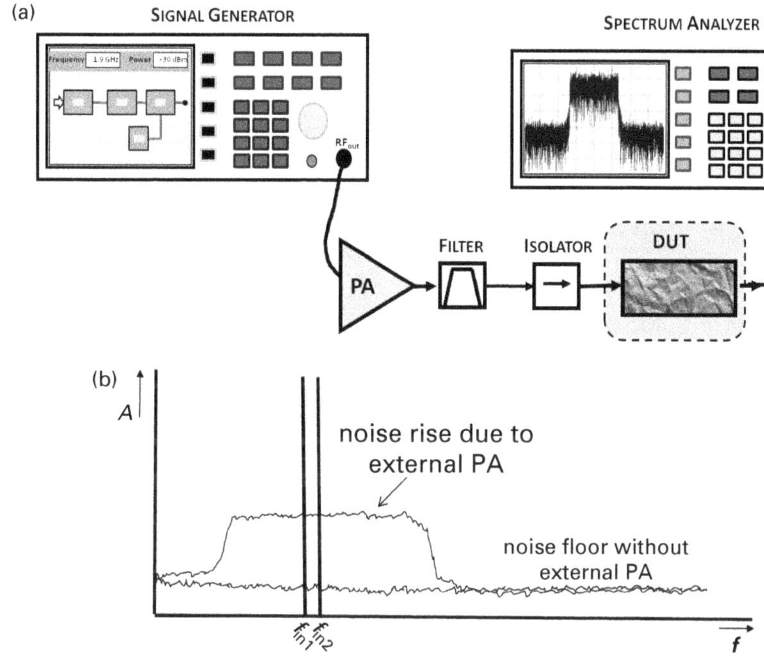


**Figure 4.7** Problems with two-tone signal generation using an AWG: (a) two-tone signal generation using an AWG and an output PA, and (b) the noise rise due to the output PA.

The main problem in this method is that of how to combine the two generator signals. One immediate solution is to use a power combiner, either resistive or not resistive. In the resistive case we will lose 3 dB, which is not an optimum scenario, since we normally need the extra 3 dB, but the resistive combiner allows us to somehow match the overall setup. Unfortunately, in this setup the signal normally traverses the combiner and, if generators are not optimally matched, it can happen that some of the output signal from generator 1 will traverse the power combiner and arrive at generator 2, creating nonlinear distortion patterns that could mask and degrade the overall measurement, as illustrated in Fig. 4.8(a). This is actually one of the most difficult scenarios to identify.

One possible way to obviate and minimize this effect is to use an isolator at the output of the generator, since the isolator will severely attenuate any return signal coming out of the power combiner. This scheme is illustrated in Fig. 4.8(b). In addition, also a filter is used at the output of the generator to minimize further any harmonic generation, and thus to clear out the generator spectra.

#### 4.2.3.3 Two-tone amplitude measurement

Using any of the previously described two-tone signal-generation mechanisms, the main goal now is to measure the nonlinear distortion generated in our DUT. Thus the two-tone signal is fed to the DUT and the output is measured using a spectrum analyzer. What can be mentioned here is that the spectrum analyzer, as presented in Chapter 2, should behave as linearly as possible, and the operator should guarantee that the input signal is within the dynamic range of the spectrum analyzer.

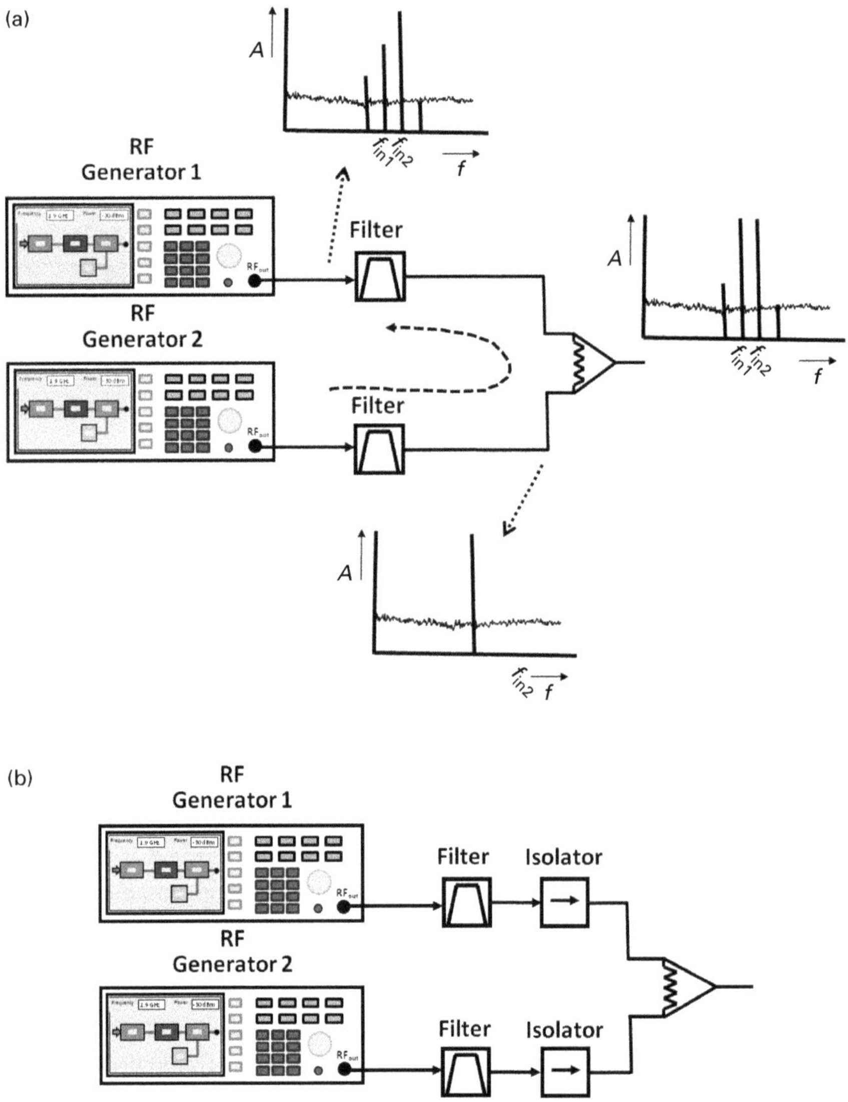


**Figure 4.8** Two-tone signal generation using CW generators: (a) two-tone distortion generation at the output of one CW generator and (b) two-tone signal generation using two CW generators.

One rule of thumb to roughly test this behavior is to change the internal input attenuation of the spectrum analyzer, and see whether the nonlinear distortion tones go up or down. Since the spectrum analyzer automatically displays the measured signal, accounting for the attenuator, the measured values should be constant unless they are being generated in the instrument itself, in which case they will change with the value of the input attenuation. If the measured values are constant, then the analyzer is measuring exclusively the input signal and not any internally generated distortion. Figure 4.9 presents a typical two-tone configuration scheme.

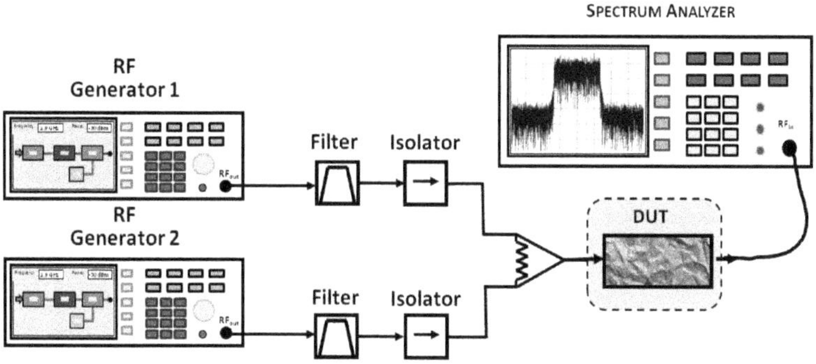


**Figure 4.9** A two-tone measurement bench, where two signals can be seen being combined in a power combiner.

Nevertheless, sometimes the nonlinear distortion generated by the DUT is so low that, in order to view it clearly, the input attenuator should be set to zero. In that case the two main tones can put the RF front end of the spectrum analyzer in a nonlinear zone, which will degrade the overall measurement. Increasing the input attenuation is not an option, since the increase in attenuation will make the noise rise, and thus mask low values of nonlinear distortion. A possibility in this case is to minimize, and ideally eliminate, the two main tones (fundamental frequencies) coming out of the DUT. That can be done with the configurations sketched in Fig. 4.10(a).

Two setups are presented for this procedure. In Fig. 4.10(a) the fundamental tones are filtered out using a diplexer filter before reaching the SA front end. The input signal for the SA is now very low, and the 0-dB-attenuator value can be used as the input attenuation, since it is expected that no distortion will be generated with this setup. This approach is effective only if the diplexer filter is of high quality, since the tones are usually close to each other, and so the filter should have a steep cut-off slope. Figure 4.10(b) presents another approach, whereby the fundamentals are eliminated by using an active cancelation approach. In this case the bench becomes more complex, but the elimination becomes easier in the case of closely spaced tones. If the tones are significantly separated from each other then the cancelation loop becomes inefficient. Owing to the delay path used, the delay should be wideband in such cases.

### 4.2.3.4 Two-tone phase measurement

Sometimes it is important to measure the phase delay between input and output tones, and for each tone itself. An example is when evaluating the impact of dynamic effects on signal degradation. In this section, we describe the solution when an NVNA is not available.

The problem of measuring phases is quite complex because, since the two sinusoids travel at different velocities (frequencies), the phase relationship is quite awkward. Nevertheless, it is possible to measure the phase change if we compare the output signal with a version of the input signal. When the tones are phase correlated, that is, if we are generating two tones using a common reference, or if they are generated using an AWG, the phase relationship can be measured using an oscilloscope with synchronous

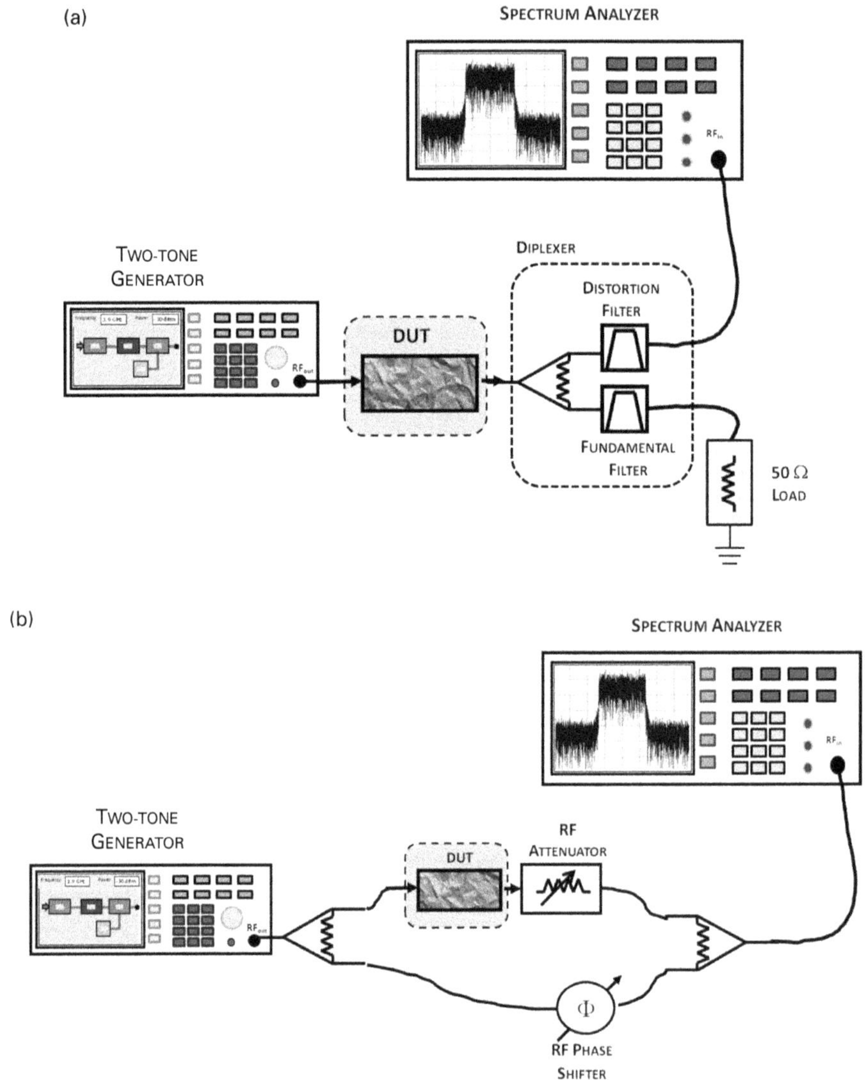


**Figure 4.10** Two-tone measurement benches for measuring extremely low values of distortion: (a) two-tone fundamental elimination by a filtering approach and (b) two-tone fundamental elimination using a cancelation-loop approach.

channels, and then the output measured signal is compared with an ideal nonlinearity created in a computer framework. This approach is illustrated in Fig. 4.11.

In this case the phase is obtained by taking a DFT of the input and output measured signals, and then the output is compared with a version of the input passed through an ideal mathematical memoryless nonlinearity, for instance $x^3$, if the IMD is the objective to be measured.

Thus, at the output of the mathematical nonlinearity and measured nonlinearity, the phase relationship is mainly a comparison of the phase difference between

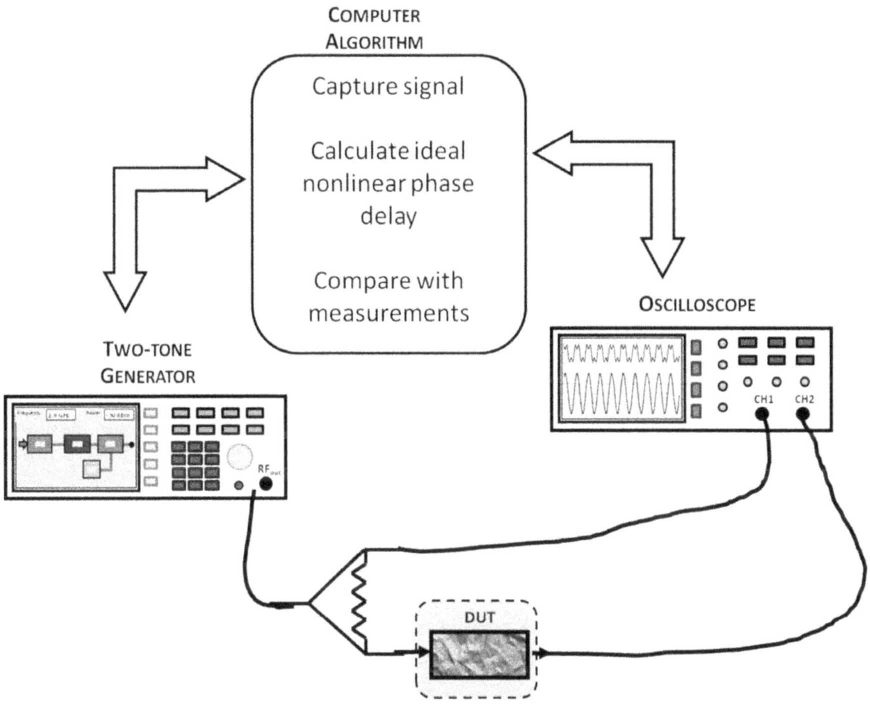


**Figure 4.11** A two-tone measurement bench for phase evaluation with correlated signals employing a known nonlinearity.

equal-frequency signals. The reader should be aware that this is valid only if the tones are correlated and if it is guaranteed that the oscilloscope measures the input signal and output signal synchronously. If this is not the case, they should be synchronized somehow [1].

There are nevertheless other approaches that are based on true measurements, rather than on computer-generated references. These methods were pursued in [2], where the approach is based on a cancelation mechanism. The basic idea of this technique is to generate a signal with the same spectral content as that of the one to be characterized, as was done in the ideal mathematical case. It is then assumed that the phase of this signal is constant whatever the amplitude of the excitation, which is typically a two-tone signal. Before starting the measurements, a calibration of the setup is mandatory. Two groups of authors [2, 3] have devoted some time to this issue. Some authors consider that the DUT is a memoryless nonlinearity, and calibrate their setup in the small-signal region of operation. In order to do this, the lower branch of the setup is made to apply the reference signal directly to the output. The two branches are then added and the result is displayed in the spectrum analyzer. The calibration process is used to cancel out the two signal branches by evaluating the output in the spectrum analyzer. When the output in the spectrum analyzer reaches a minimum, the signals being compared have opposite phases. At this time the vector modulator is measured, and this phase will be used for future calibrations. Some authors consider that the IMD phase measured

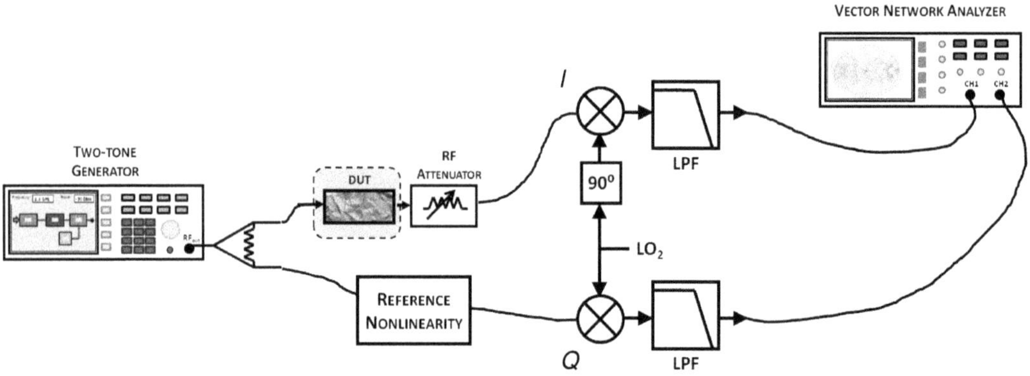


**Figure 4.12** A two-tone measurement bench for phase evaluation employing a known nonlinearity.

in the small-signal case is equal to the phase of the carrier frequency, whereas others consider that the carrier frequency is of the opposite phase. But in real DUTs, which present dynamic effects, this is no longer true, and other schemes are necessary.

The measurement procedure is continued by returning the input signal to its default value, and the output is again eliminated by the output cancelation setup. The actual phase value is the difference between the value measured in the vector modulator and the calibration value.

Another alternative method by which to measure the IMD phase in a continuous and automatic way is to follow similar ideas, but now considering a known reference. In this case, the input signal will be passed through a reference nonlinearity and compared with the output signal for the same frequency component, as illustrated in Fig. 4.12.

This setup assumes that the input signal is first divided into a power splitter and then fed to the DUT and to a known reference. The reference can be any known memoryless nonlinear device, for instance a mixer or a high-frequency amplifier. At the output the signals are down-converted and filtered out, so a VNA can be used for measuring both signals, which is possible since they are at the same frequency, and thus one can measure the phase difference between them. More information on this setup can be found in [4].

#### 4.2.3.5 Two-tone measurements in the presence of dynamic effects

As introduced in Chapter 1, certain nonlinear DUTs present what is called memory effects [5]. The way to measure dynamic intermodulation effects is by using a two-tone signal and varying the spacing between the tones, as shown in Fig. 4.13. The bench used for this evaluation is the same as the one used in Section 4.2.3.3, but now the tone separation should be changed accordingly. The tone separation can be done manually, point by point, or chirp signal generation can also be used, as presented in Section 3.5, with extra care regarding the time window used to capture the nonlinear distortion.

### 4.2.4 VNA measurements

The primary use of a VNA is obviously for performing *S*-parameter measurements, as will be outlined in Section 4.2.4.1. However, modern high-end VNAs have

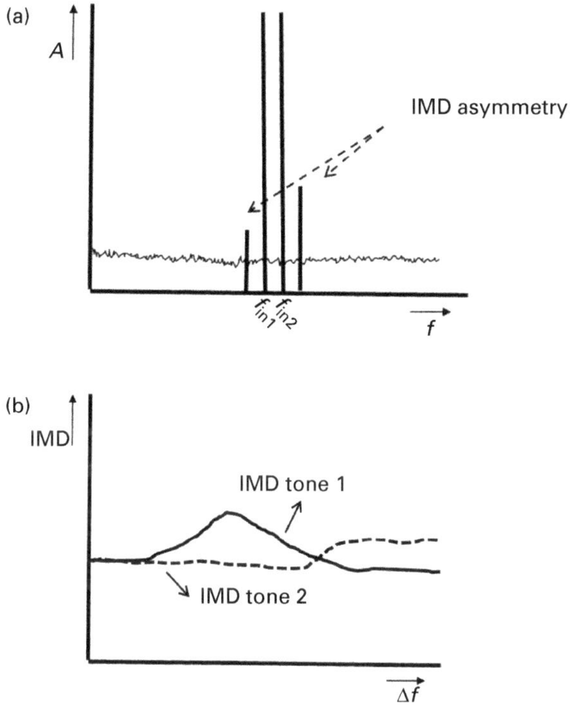


**Figure 4.13** The impact of two-tone nonlinear dynamics: (a) two-tone IMD measurement when memory effects are visible and (b) two-tone IMD variation with tone spacing.

extensive capabilities, and are in fact able to measure also the parameters discussed above (power, noise figure, two-tone measurements). This aspect will be elaborated on in Section 4.2.4.2.

### 4.2.4.1 The procedure for *S*-parameter measurements

Today's VNAs have built-in computers with extensive GUIs that guide the user through the measurement procedure. Nevertheless, it is important to have a good understanding of the various steps, illustrated in Fig. 4.14, in order to ensure that one will obtain accurate measurements in the end.

Before the calibration can start, the user has to insert or confirm a number of settings. An important setting is the frequency range (i.e., start and stop frequencies, number of measurement points, linear or logarithmic sweep). Since this can no longer be changed once the calibration has been done, this setting must be considered carefully at the beginning. The more measurement points, the finer the resolution of the data, but also the longer the measurement will take. A dense grid is fine for a passive component, but, as soon as one has to vary operating conditions such as the DC bias, one easily ends up with overnight measurements if too many frequency points are requested.

The next parameter is the power range. The higher the incident power, the better the signal-to-noise ratio, and hence also the better the measurement. But, if the incident power is too high, the DUT may start behaving nonlinearly, and then the measured *S*-parameters are not correct. A way to check for this is to calibrate the system at a

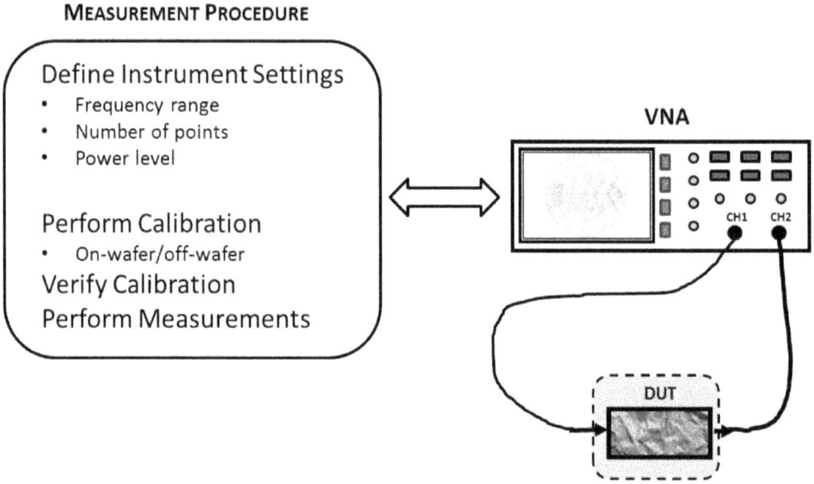


**Figure 4.14** The VNA measurement sequence.

nominal power, −20 dBm for example, and then manually vary the incident power and examine whether the value of $S_{21}$ changes. Depending on the instrument, the user can change the RF source power and/or the instrument may have a bank of internal attenuators. When ramping up the power, the power at which $S_{21}$ starts to change, usually by decreasing, is the maximum power that can be set. When you are measuring over a wide frequency range, it is advisable to introduce a power slope to compensate for the frequency-dependent loss in the cables (and probes) between the instrument and the DUT.

The power levels going into the instrument should be checked as well. If the DUT has a high gain and/or high output power, the internal attenuator at port 2 (and possibly that at port 1 as well) should be activated in order to avoid having the instrument driven into nonlinear operation, or even worse, being damaged. It is a trade-off exercise, because the additional attenuation reduces the dynamic range, and thus the quality of the measurement.

The dynamic range can be improved by increasing the averaging and reducing the IF bandwidth, but this is a trade-off as well due to the increase in measurement time entailed.

Finally, note that the actual settings are manufacturer-dependent, so check the manual of your instrument to ensure that all necessary settings have been set.

Once the settings have been set, the calibration procedure as explained in Section 2.6 can be performed. In practical cases, the user has to select the calibration approach, and then the VNA GUI will guide the user in connecting the various standards in the case of a mechanical (connectorized or on-wafer) calibration. In the case of an e-calibration, the user has to connect the e-calibration module only once and then the instrument software executes the calibration procedure automatically.

At the end of the calibration, it is a good habit to validate the calibration. This can be done by performing a measurement on a standard that was not part of the calibration

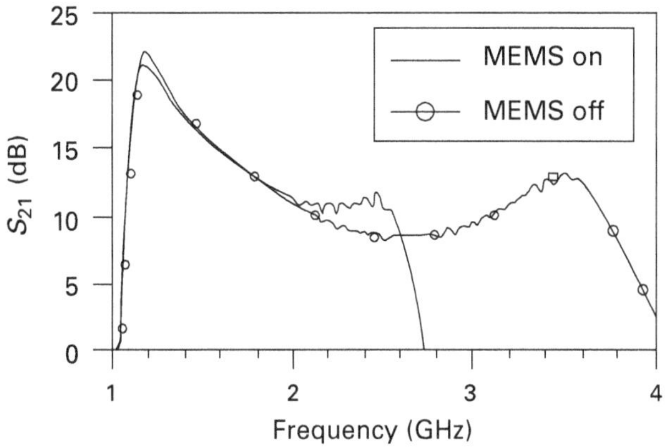


**Figure 4.15** Measured $S_{21}$ of a reconfigurable multi-band amplifier. The amplifier is switched between two bands by putting the MEMS switch on or off [6]. © IEEE.

procedure. Alternatively, make use of a passive component for which you know the characteristics to expect. If both options are not available, you may remeasure a standard that was part of the calibration procedure. In this case, you are checking measurement repeatability rather than validating the calibration. But if the result deviates strongly from the expected value, you can be sure that something went wrong during the calibration procedure. For example, if you connect a short or open standard to port 1, the magnitude of $S_{11}$ should be close to 0 dB, to within a few tenths of a dB.

Once the calibration has been validated, you can proceed to the actual measurement. Measurements of the $S$-parameters are usually the first measurements conducted on a new microwave circuit, after the DC functional testing has turned out to be correct. The gain and matching are measured, and compared with the circuit's datasheet or simulated specifications. When the results are in agreement, more complex testing such as two-tone measurements and noise-figure measurements may follow next. Figure 4.15 illustrates the measured $S_{21}$ of a reconfigurable multi-band amplifier [6].

#### 4.2.4.2 Extended measurements

As explained in Section 2.6.2, the standard VNA calibration assumes that the measured quantities are ratios. For this reason, the magnitude and phase of the normalization coefficient in Eq. (2.34) is not determined. It is, however, straightforward to calibrate the magnitude using a power meter, and therefore this utility is built in as standard in today's VNAs. Note that determining the phase of the normalization coefficient is more complex and necessitates a full-blown NVNA.

A distinction is made between source and receiver power calibrations. The source power calibration calibrates the power level of the RF source that is used as the excitation in the measurements. If the excitation is to be applied at port 1 of the DUT, the power sensor is connected to port 1. The VNA has a built-in algorithm to sweep the source across the set frequency range and to adjust the RF source power level until the target power level gets read by the power sensor.

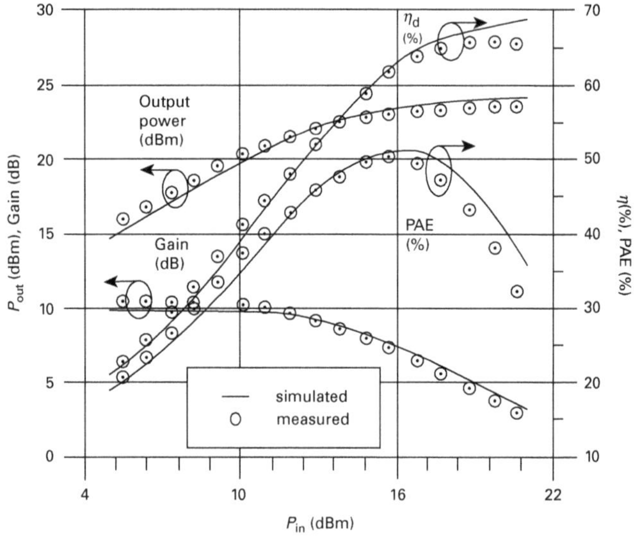


**Figure 4.16** Power amplifier results based on single-tone power measurements using a VNA [7]. © IEEE.

The receiver power calibration mathematically removes frequency-response errors in the specified VNA receiver. The readings are adjusted to the same value as the source power calibration level (or to a specified offset value). The first step in this procedure is to perform a source power calibration. Next, a "through" is to be connected between the source port and the receiver. The built-in algorithm then automatically determines the correction to be applied.

After the power calibration, power measurements become possible. Combined with a DC measurement, figures of merit such as gain, output power, drain efficiency, and PAE can be characterized, as illustrated in Fig. 4.16 [7].

Since modern VNAs often have two, or even four, internal RF sources, also two-tone measurements are straightforward to perform. This provides an alternative to the procedures explained in Sections 4.2.1 and 4.2.3.

Also mixer measurements can be performed with a VNA. The combined set of $S$-parameter measurements and power calibration is sufficient to obtain scalar mixer measurements, meaning magnitude only. To determine also the phase response and group delay, or in other words carry out vector mixer measurements, one usually makes use of a reciprocal mixer–filter pair during the calibration [8]. This procedure is quite cumbersome, and therefore we refer the reader to the recent approach making use of the NVNA calibration concept [9]. In this approach, a reciprocal mixer–filter pair is no longer required.

It is even possible to perform noise-figure measurements with a VNA. Depending on the manufacturer and frequency range, dedicated noise receivers may be implemented in the instrument. In the following, we assume that the standard receiver is used as a noise receiver. Obviously, the measurement accuracy will be less good than with a dedicated noise-receiver measurement setup, such as the one described in Section 4.2.2.

The calibration procedure starts with a source power calibration, as explained above. Next, a "through" connection is made between the calibrated source port and the receiver that is being used as a noise receiver. The instrument's built-in software characterizes then the gain bandwidth and noise level of this receiver. Next, a regular $S$-parameter calibration is performed. An optional step is to correct for the mismatch of the source port, in which case the accuracy will be better because noise-figure measurements assume that the source impedance is 50 Ω.

## 4.2.5 NVNA measurements

The NVNA measurement section is split into two parts. First, the general measurement procedure is described, together with practical examples. Next, the application of load-pull measurements using an NVNA is elaborated in more detail.

### 4.2.5.1 The measurement procedure

The measurement procedure is outlined in Fig. 4.17. Compared with the VNA measurement procedure, the phase of instrument settings is shorter and is usually combined with the calibration. NVNA measurements have a wide range of degrees of freedom for experimental settings (power sweep, frequency sweep, single-tone, two-tone, or multisine excitation, load-pull, ...), but this can be decided upon after calibration.

Since the mixer-based NVNA is based on the VNA architecture, the notes on calibration in Section 4.2.4.1 are largely applicable in this case as well. One difference is that the $S$-parameter calibration is performed at the highest power level that still ensures receiver linearity, so the DUT power levels do not have to be considered yet. This is possible since a power calibration is always part of the NVNA calibration procedure.

In terms of initial settings, also here it is important to think carefully about the frequency range prior to calibration. The largest common divider of the frequencies to be characterized in the intended measurements is to be taken as the calibration frequency. In the case of the NVNA, the default of 10 MHz is often selected for reasons

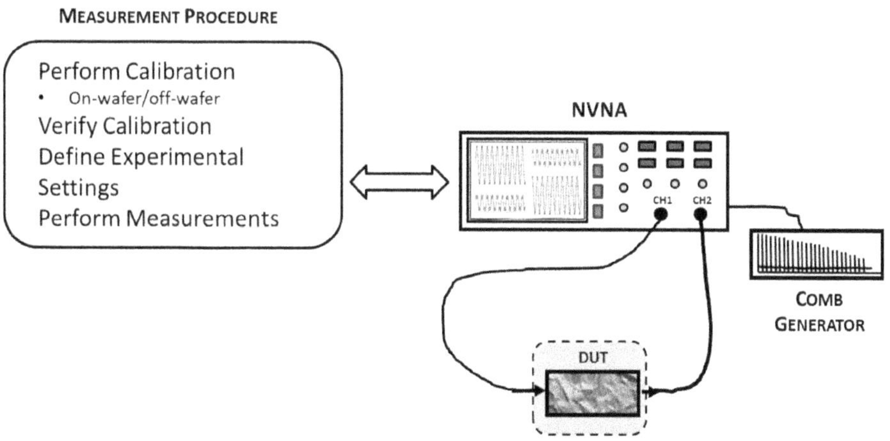


**Figure 4.17** The NVNA measurement procedure.

of flexibility. The older sampler-based LSNA architecture was more limited, in that the calibration frequency had to be between 600 MHz and 1.2 GHz.

As explained in Section 2.7.2, the calibration procedure consists of three parts: linear calibration, power calibration, and harmonic phase calibration. The procedure is automated in today's commercial instruments. The software informs the user where to connect which standards.

Since the instrument is relatively recent, there exists as yet no calibration verification element. So the recommendation is to first check the *S*-parameter calibration verification, and then measure a device regarding which one knows which characteristics to expect. Since this device is an in-house-selected device, one cannot be sure of its accuracy but at least it allows one to detect major calibration issues.

The output data can be visualized (and stored) in both frequency- and time-domain formats. Both formats can be useful for behavioral modeling (see examples in Section 4.3) and to check for modeling accuracy. The frequency-domain results can give information similar to that from single-tone power-meter measurements (see Section 4.2.1), and from two-tone and multi-sine excitation-based characterizations (see Sections 4.2.3 and 4.2.6).

Figure 4.18 visualizes a time-domain result. The measurements were obtained on the sampler-based architecture with low-frequency (LF) extension [10]. The latter enables one to characterize the baseband response jointly with the response at RF. Examining the RF only (Fig. 4.18(a)) may lead to incorrect interpretation, such as that the device is not pinching off. The correct physical behavior, combining the baseband and RF responses, is depicted in Fig. 4.18(b).

The added value of using an NVNA rather than a VNA is that the response around the harmonic frequencies can be characterized as well. The added value of using an NVNA rather than a spectrum analyzer is that both the amplitude and the phase of the spectral components can be characterized, such that the time-domain representation becomes possible. These two aspects are jointly illustrated in Fig. 4.19, which depicts the magnitude and phase of the complex envelopes around the carrier frequency and around the second-harmonic frequency versus time. The DUT is a packaged amplifier designed for 4.9-GHz wireless applications. The excitation is a 63-tone multi-sine designed to mimic

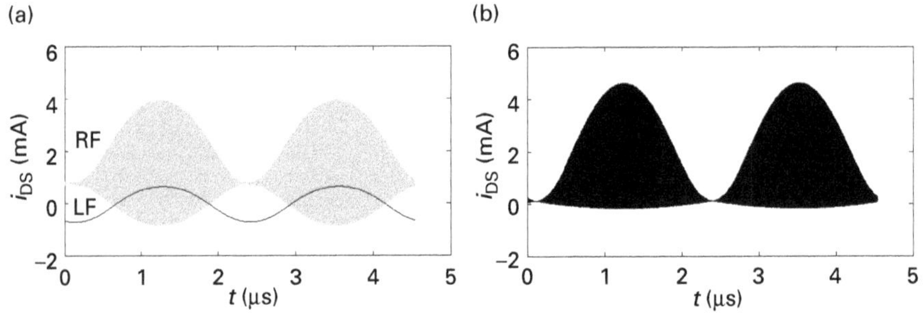


**Figure 4.18** Measured time-domain waveforms of a FinFET under two-tone excitation [11]: (a) baseband (LF) and RF responses and (b) combined baseband and RF response. © IEEE.

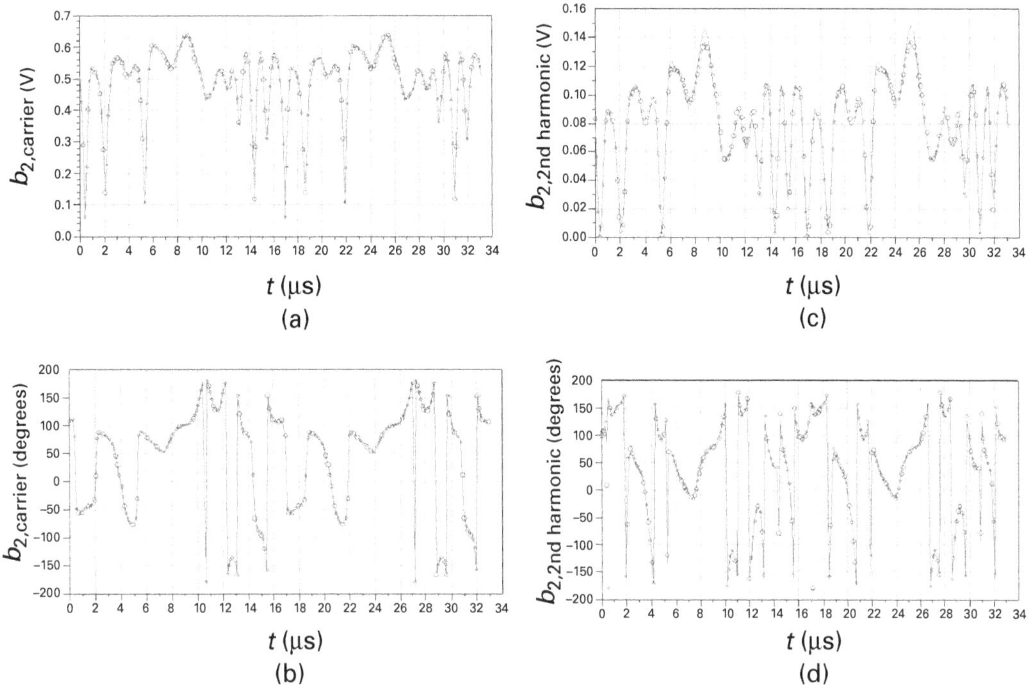


**Figure 4.19** Measured (circles) and simulated (solid trace) waveforms in (a) magnitude and (b) phase of the complex envelope around the RF carrier frequency, and of (c) magnitude and (d) phase of the complex envelope around the second RF harmonic. The DUT is an amplifier and the excitation is a multi-sine mimicking QPSK modulation [12]. © IEEE.

a 1.6-MHz-bandwidth QPSK-modulated digital signal [12]. Figure 4.19 is comparing the measurements against a behavioral model [13], so it is also an example of model validation.

An application of high interest is extending the NVNA setup to enable load-pull measurements, which will be discussed in the next section.

### 4.2.5.2 Load-pull measurements

Load-pull is the term applied to the process of systematically varying the load impedance presented to the DUT, which is most often a transistor, to assess its performance as a function of this load impedance. Source-pull exists as well, in which case the impedance presented to the DUT at its input is being varied. The most common application of source-pull is characterization of the noise parameters of transistors, while load-pull is inherently connected to the design of power amplifiers. The load can be varied in a passive or active way. Both approaches are illustrated in Fig. 4.20 [14].

The passive approach makes use of a mechanical tuner to vary the impedances. Some examples are depicted in Fig. 4.21. The drawback of mechanical tuners is that the magnitude of the realized reflection coefficient has a practical upper limit due to losses in cabling, and probe if on-wafer, between the tuner and the DUT. So no impedances close

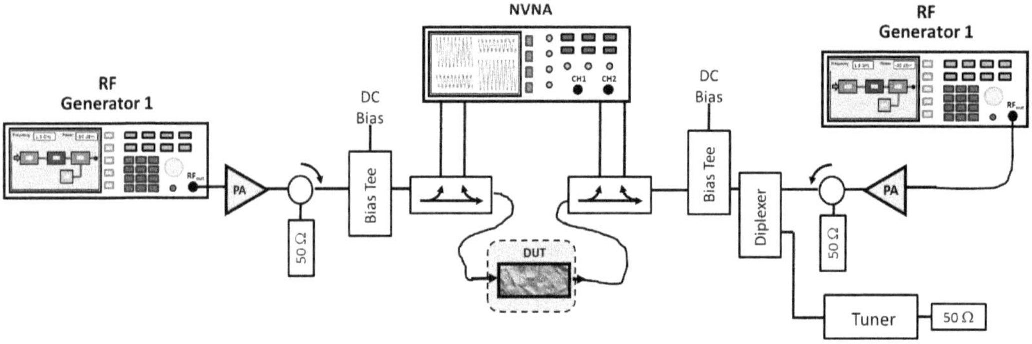


**Figure 4.20** The measurement bench of an NVNA extended with harmonic load-pull capability [14].

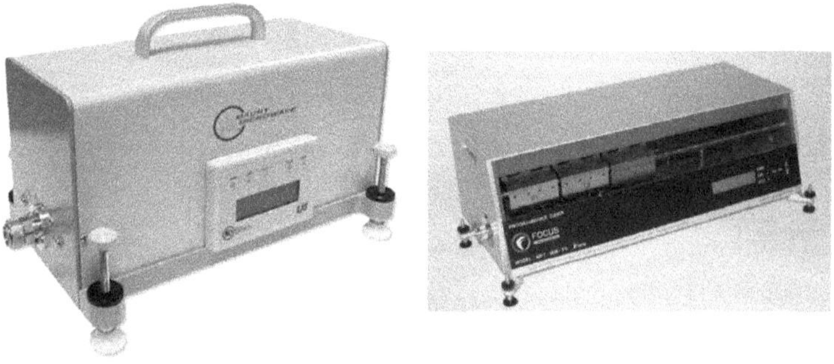

**Figure 4.21** Two mechanical tuners. © Maury and Focus.

to the edge of the Smith chart can be realized. In the active approach, a signal is injected at port 2 of the DUT. Both closed-loop and open-loop configurations exist. Figure 4.20 shows an open-loop implementation. The signal generated by RF source 2 creates an incident wave of amplitude $a_2$ going into the DUT's output. The obtained reflection coefficient is $a_2/b_2$, with $b_2$ the amplitude of the scattered wave at the DUT's output, port 2. If $a_2$ is larger than $b_2$, instabilities may occur, and this is an important drawback of the active-load-pull method. Impedances at the edge of the Smith chart can be achieved by properly choosing the amplitude of $a_2$. This may require an additional power amplifier at port 2, as shown in Fig. 4.20, in order to generate adequate power relative to $b_2$. Often passive and active load-pull are combined in such a way as to optimize the coverage of the Smith chart while keeping the potential instability problem under control. In such cases, the passive tuner is set to a load near the expected optimum load of the DUT, and excursions using active load-pull are made around this load set by the passive tuner. Figure 4.20 depicts another way of combining passive and active load-pull. In this example, the tuner is used for load-pull at the fundamental frequency, and the active injection is employed for the purpose of load-pull at the second-harmonic frequency. These two types of load-pull are combined using a diplexer. For

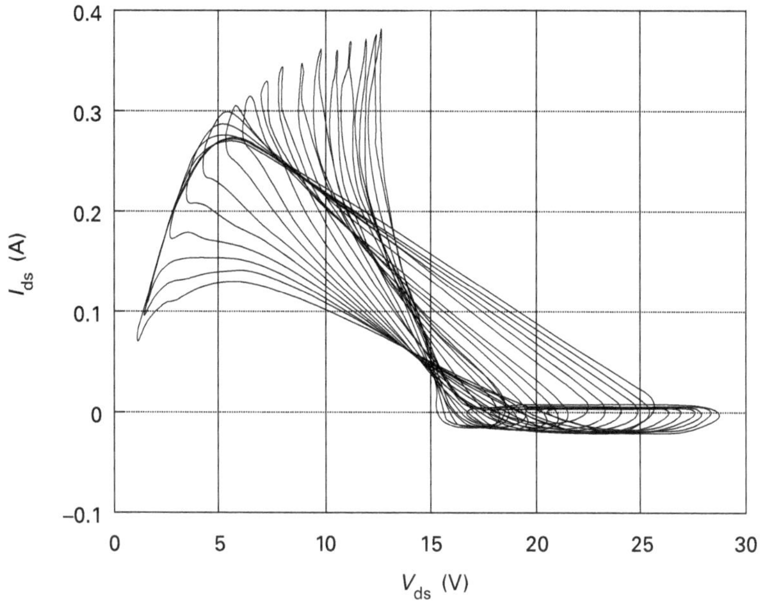


**Figure 4.22** Dynamic load-lines under various load conditions measured on a GaN HEMT [14]. © IEEE.

completeness, it should be noted that harmonic tuning can also be achieved just using mechanical tuners that have been specifically designed for this purpose.

The setup in Fig. 4.20 combines the load-pull capability with an NVNA. Traditionally, load-pull has been combined with a power meter or a spectrum analyzer. This is adequate when the aim is to obtain figures of merit such as output power or power-added efficiency as functions of the load impedance. The added value of the NVNA is that dynamic load-lines can be straightforwardly obtained and analyzed. An example is shown in Fig. 4.22. The dynamic load-line means that the time-domain waveform of the current at port 2 is plotted against the time-domain waveform of the voltage at port 2. The load-line changes in slope and shape as a function of the (harmonic) load realized. Such dynamic load-lines are highly useful in power-amplifier design. By shaping the dynamic load-line, the amplifier's performance can be optimized. This way of designing amplifiers is called "waveform engineering" [15, 16].

### 4.2.6 Modulated signal measurements

Modulated signals or multi-sine measurement benches are gaining increased importance due to growth of wireless communication systems. In contrast to the other benches described in the preceding sections, these benches are always based on an AWG generator. The reason is that the signal should mimic as closely as possible real communication environments, and thus the generation of spectrum masks is fundamental for a correct evaluation of the DUT.

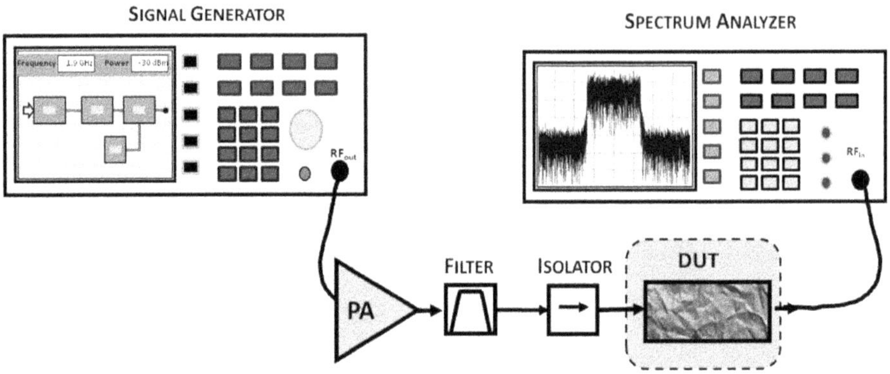


**Figure 4.23** The typical configuration for a modulated signal measurement evaluation.

As explained in Chapter 1, the important figures of merit in connection with modulated signals are the ACPR, NPR, CCPR, EVM, etc., all of which impose the use of spectrum analyzers or VSAs, which could be used in conjunction. It is our view that a VSA, or for certain applications an RTSA (when variation over time is important), would be the appropriate instrument to use in this case. In the case of multi-sine excitations, also the use of an NVNA may be considered. As explained in Section 2.7, an NVNA can, however, not be adopted if the spectrum is continuous, as in the case of realistic modulated signals.

Figure 4.23 presents the typical configuration (which is exactly equal to the AWG two-tone bench).

In this case the input signal synthesized in the AWG passes through the DUT and then the output signal is measured by a VSA. As explained earlier, similar care should be taken with respect to the RF front end of the VSA in order not to saturate it.

#### 4.2.6.1 Adjacent-channel power measurements

The first important measurement to be done in modulated signal evaluation is to account for the amount of spectral regrowth that a nonlinear device can generate. That can be done by generating a signal pattern in the AWG, which should be as close as possible to the wireless system standard to which the DUT will be subjected, and feeding it with that signal. The output will then be measured using a spectrum analyzer, as in Section 1.5.3. In the case of $ACPR_T$ measurements, either $ACPR_L$ or $ACPR_U$, the measurement is made in the SA by using a convenient mask. This means that the fundamental signal should have a mask and the spectrum regrowth should have another mask. The power in the fundamental mask is evaluated as the power inside the limits, and the calculations are done accordingly. Figure 4.24 presents typical output-distorted spectra.

As can be seen in Fig. 4.24(a), it is very difficult to define precisely the masks in a truly modulated signal, since most modulated signals have some kind of spectral roll-off, and thus the precise definition of where the ACPR mask starts becomes a cumbersome problem. That problem can be obviated using a multi-sine signal as seen in Fig. 4.24(b), if the multi-sine presents statistical values similar to those discussed in

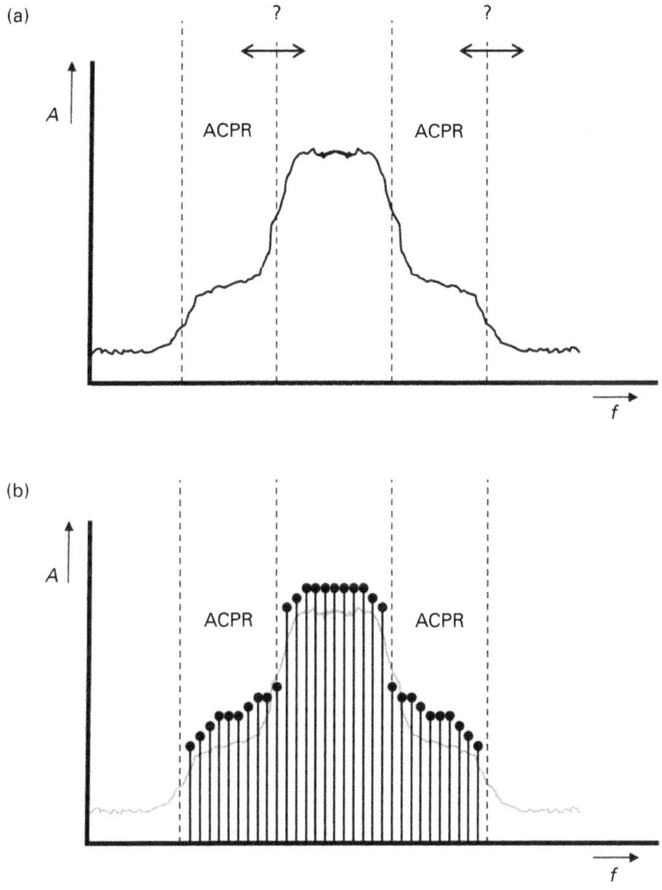


**Figure 4.24** ACPR measurement: (a) typical ACPR in continuous spectra and (b) typical ACPR in multi-sine spectra.

Section 1.5.3. Nevertheless, the evaluation of the ACPR in signals with continuous spectra can be obtained with great confidence if, instead of $ACPR_T$, we use the spot ACPR, $ACPR_{SP}$, as explained in Section 1.5.3.2. In this case the mask for the fundamental power measurement can be the same as used before, but the mask for the distortion is, in principle, far away from the roll-off, and thus perfectly identifiable and clear, as illustrated in Fig. 4.25.

It should also be stated here that some SAs already have a built-in capability to measure these figures of merit automatically, imposing only that the operator define the corresponding mask limits.

### 4.2.6.2 Noise-power-ratio measurements

As presented in Section 1.5.3.3, the ACPR is not a complete measurement approach since it evaluates mainly nonlinear distortion appearing in adjacent-channel frequencies. Nevertheless, the most important aspect for a modulated signal is its co-channel

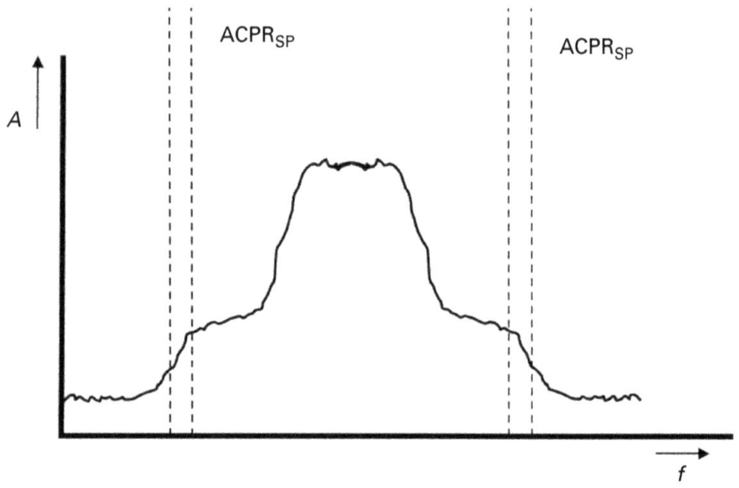


**Figure 4.25** ACPR spot measurement, where the spot bandwidth is identified.

frequencies, since those will decrease the SNR and thus will degrade the BER. Thus measuring co-channel distortion is crucial. The first co-channel evaluation is that of the so-called noise power ratio, NPR. The NPR is measured, as explained in Section 1.5.3.3, as the distortion appearing in a notch that is made in the input signal. The input signal to be used could be a multi-sine or a truly modulated signal, but a spectrum notch should be made prior to exciting the device.

If a multi-sine can be used then the notch is easily implemented by switching off the middle terms. This should be done with care, since in an AWG the middle term normally falls on top of the local oscillator, and thus some LO leakage may appear in the generated spectrum. The bench operator should be aware of this fact. The same happens with a modulated signal, but then the signal should be carefully designed in a digital way prior to exciting the DUT. The proposed scheme for this design is to create a modulated pattern in baseband and then, after filtering using a high-pass filter, the signal can be $I/Q$ up-converted. Again the up-conversion should be done with care in order to avoid the appearance of the LO at the output. Figure 4.26(a) presents these signals.

The measurement is then done using a VSA or an SA. The power is measured in the fundamental-frequency mask and in the notch mask. Again, if a multi-sine is used, the notch power is easily measured, but in the case of the continuous spectrum the roll-off of the filter can impose some errors that should be accounted for.

### 4.2.6.3 Co-channel power-ratio measurements

Although NPR measurements can be very well understood and can be a great help for RF designers, it can be proven [17] that NPR measurements do not capture the overall impact of nonlinear devices excited by a modulated signal. The main reason for this is that the input signal was changed, and thus the response will be different from the response of the real signal. Thus some authors have proposed the use of a more robust measurement procedure for identifying co-channel distortion, as presented in Section 1.5.3.3. This figure of merit is called the co-channel power ratio (CCPR). It is

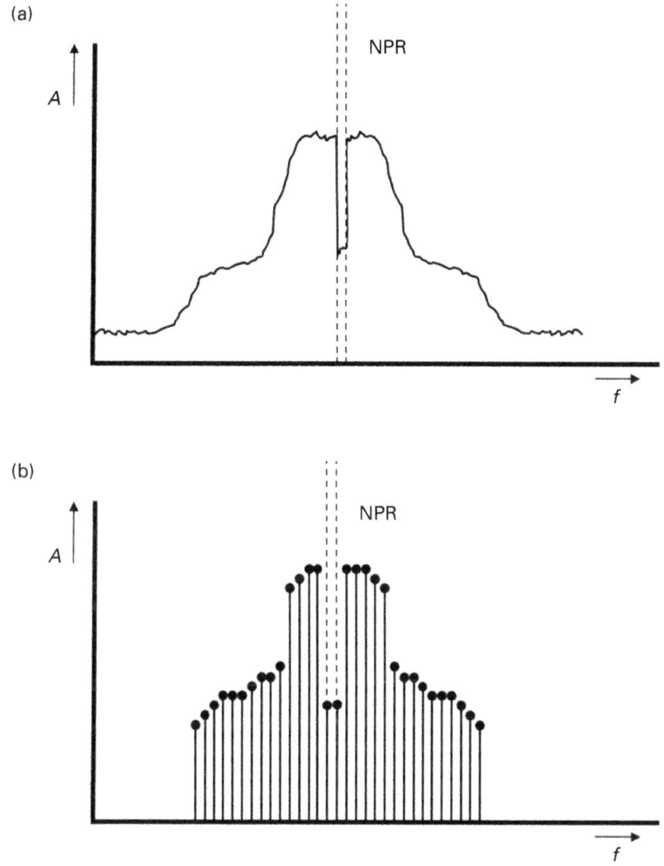


**Figure 4.26** NPR signal generation: (a) a typical NPR measured signal for a continuous spectrum and (b) a typical NPR measured signal for a multi-sine spectrum.

measured using a setup similar to the nonlinear-distortion-mitigation mechanism. The main idea here is to measure the real nonlinear distortion noise, which means the nonlinear distortion appearing at the co-channel frequencies that are not correlated with the fundamental signals. But, to achieve this, especially since they are on top of each other, it is important to eliminate the fundamental correlated signal prior to the measurement. A possible way to do that is sketched in Fig. 4.27.

The setup is composed of an AWG followed by a bridge setup inspired by the cancelation loop of a feed-forward linearizer system [17], and finally an SA or a VSA. The upper branch contains the DUT, while in the lower branch a simple phase-delay device is used to mimic the linear behavior of the DUT. The auxiliary branch (lower branch) can be deactivated by switching the power divider and combiner ports to matched loads, for complete fundamental and distortion output readings. The final power combiner is used as the signal-subtraction component, and the output is measured by the SA. The setup includes also two attenuators, one at the input, to control the excitation level, and another one in the upper band after the DUT, which, combined with the phase delay in the lower branch, mimics the BLA of this DUT.

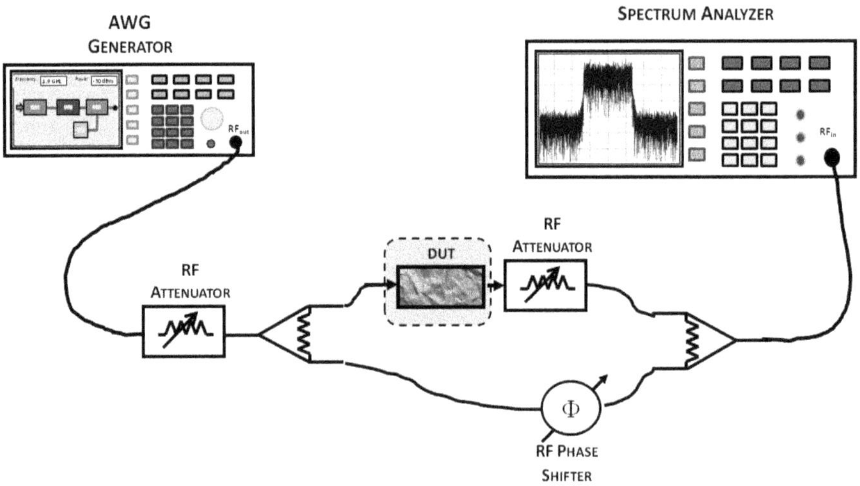


**Figure 4.27** A typical CCPR measurement bench.

In order to understand this procedure, consider that the output of the DUT is composed of a linear replica of the input and a nonlinear distortion: $X_{\text{out}} = G_{\text{BLA}}X_{\text{in}} + D$, where $G_{\text{BLA}}$ is the best linear approximant and $D$ is the nonlinear distortion.

If we have the DUT operating with a low-power signal excitation, then the output will be mainly a linear replica of the input. In this case the loop can be tuned to eliminate the linear part, and the phase delay combined with the action of the second attenuator can be changed in order to observe only noise in the SA, as illustrated by Fig. 4.28(a). In this state the linear contribution is completely eliminated even when the input excitation signal is increased. Since the linear components in the two branches are similar in both branches, this is a procedure to identify the linear gain $G$. When the input signal power is increased, and the input attenuation decreased, the output will be $X_{\text{SA}} = G_{\text{BLA}}X_{\text{in}} + D - GX_{\text{in}}$.

The measured value in the SA will account for the overall distortion in the nonlinear device. This means that it will account for both correlated and uncorrelated distortions, which constitute in fact the CCPR to be measured.

If just the uncorrelated co-channel distortion is to be measured, it can also be obtained using this setup. In that case the setup should be tuned when the device is excited at its full power. In this case the components that can be subtracted are just the correlated ones, since the uncorrelated distortion cannot be mimicked by the lower branch. This idea was initially proposed in [18] for measuring nonlinear co-channel distortion. The result of such a measurement can be seen in Fig. 4.28(b).

### 4.2.6.4 Modulated information

Up to now the instrumentation benches we have focused on were applied mainly to measure signals at RF. But, as explained in Chapter 1, RF signals normally contain

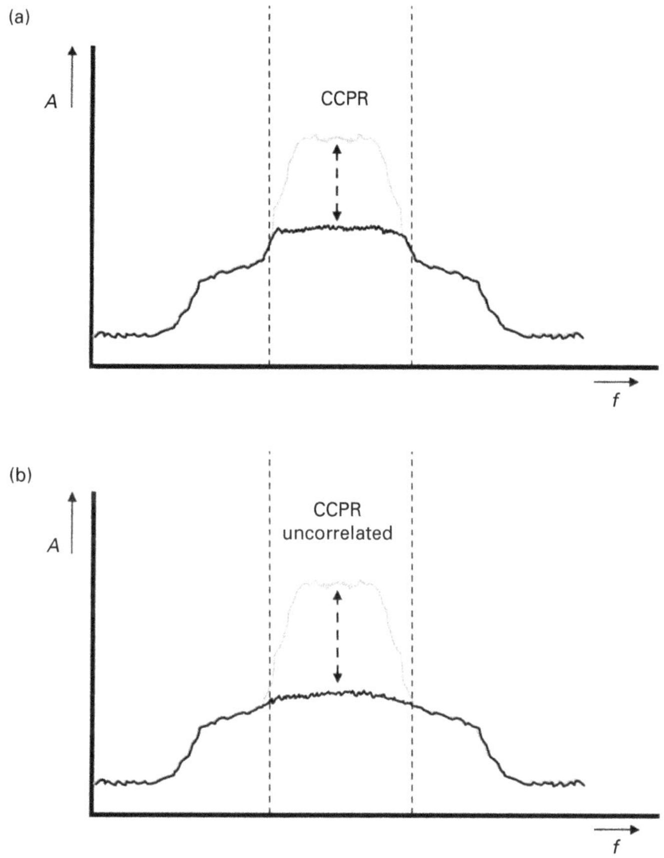


**Figure 4.28** A CCPR measurement: (a) a typical CCPR canceled signal, when the system is tuned at small signal condition; and (b) a typical uncorrelated CCPR measurement, when the system is tuned at large signal condition.

information that is modulated either in phase or in amplitude. In this way, signals can transport information and thus allow communication between several points in a wireless network.

Despite the fact that RF figures of merit give a nice perspective on the overall performance of the system, they do not give direct readings about the information being transmitted. This information is usually found by evaluating real modulated signals, either after demodulation or at the envelope layer. That is why the evaluation of the $I/Q$ time evolution is very important, since it allows one to evaluate the information it contains.

One figure of merit that it is important to measure in this context is the EVM, which was explained in Section 1.6.2. The EVM gives us information about the degradation of the constellation diagram of a digital modulated signal traversing non-ideal scenarios,

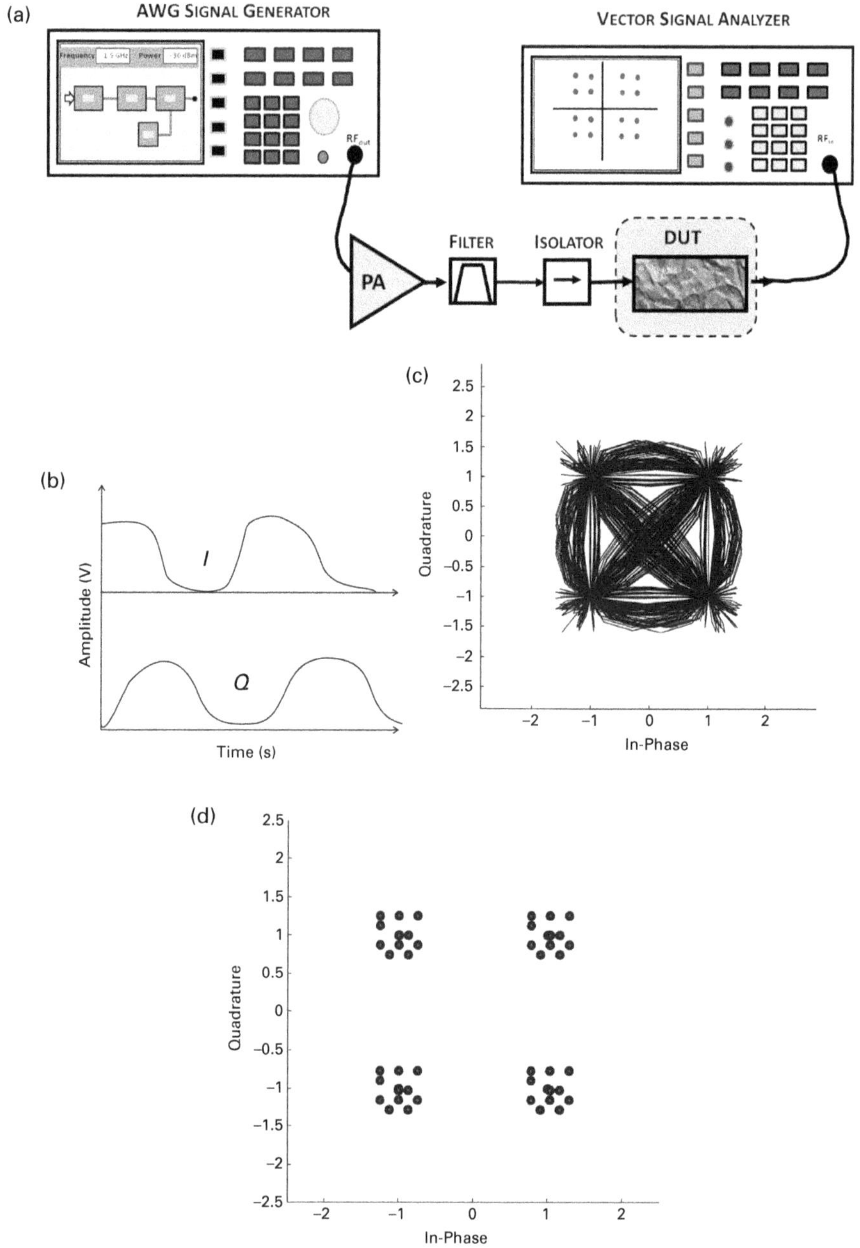


**Figure 4.29** Baseband measurements: (a) a VSA measurement bench, (b) a typical $I/Q$ measurement, (c) a typical $I/Q$ constellation, and (d) a typical EVM measurement.

be they linear or nonlinear. If the objective is to capture these behaviors in the laboratory, then a VSA, as covered in Section 2.4, should be used, since it allows the $I/Q$ signal to be captured over time, and then further it can be evaluated over time. The bench and a typical measurement of the $I/Q$ signals can be seen in Fig. 4.29.

The measurements when performed in terms of $I/Q$ are done at the complex envelope level, and thus capture mainly in-band information. It is possible with this VSA approach to measure also the ACPR, NPR, or CCPR values presented previously if the $I/Q$ signal is converted to the frequency domain. But, more importantly, the main quantity that can be measured is the deviation of the $I/Q$ signal constellation from the original or transmitted one, as shown in Fig. 4.29(d).

If the instrument has the demodulation algorithm installed, it is also possible to demodulate the signal completely and to evaluate the BER of the overall system. In order to enable this, the receiver (VSA) should have knowledge of the transmitted sequences. This is normally done using predetermined pseudo-random bit sequences, normally coded as PRBS, which are known both at the transmitter, namely an AWG, and at the receiver.

#### 4.2.6.5 Time-division signal measurements

New wireless technologies are imposing stronger challenges in terms of their instrumentation and characterization needs. One of those limitations is related to the change of the signal waveform over time, which is mainly due to the use of such advanced configurations as time-division multiple access (TDMA) and frequency hopping (FH), or more complex mechanisms for wireless systems such as the one used in WiFi systems like carrier sense multiple access (CSMA). All these new systems impose a behavior over time that should be carefully characterized and evaluated.

Similar figures of merit to those used for traditional characterizations can continue to be used here, but now most of them will have a variation over time. For instance, we could define something as the ACPR over time ACPR($\tau$), which gives information as an inherent variation with time. Thus the equipment presented in Section 2.5, namely the RTSA, becomes a key point in these scenarios.

In this case the setup for evaluating these parameters is mainly achieved by connecting the DUT to the RTSA, and then measuring the variation of the spectra over time. Again it should be noticed that what we are measuring here is an $I/Q$ signal that varies with time, which entails removing from it the carrier information. Figure 4.30(a) presents this configuration.

In Fig. 4.30(b) an example measurement can be seen. This bench is actually very important for wireless regulators, since it allows them to monitor the spectrum occupancy, and to identify variations over time, as well as to create triggers that could be activated when the spectrum is corrupted in a specific area. Developments are being made in order to use this type of instrumentation in the new concepts of cognitive radio and white-spectra occupancy. The reader can browse [19] for more information. Again in this measurement it is very important to select the correct time window for the spectrum water-fall evaluation, as discussed in Section 2.5.2.

### 4.2.7 Mixed-signal (analog and digital) measurements

With the new advances in software-defined radio (SDR), the wireless community is starting to have other special needs when it comes to evaluating DUTs. In an SDR

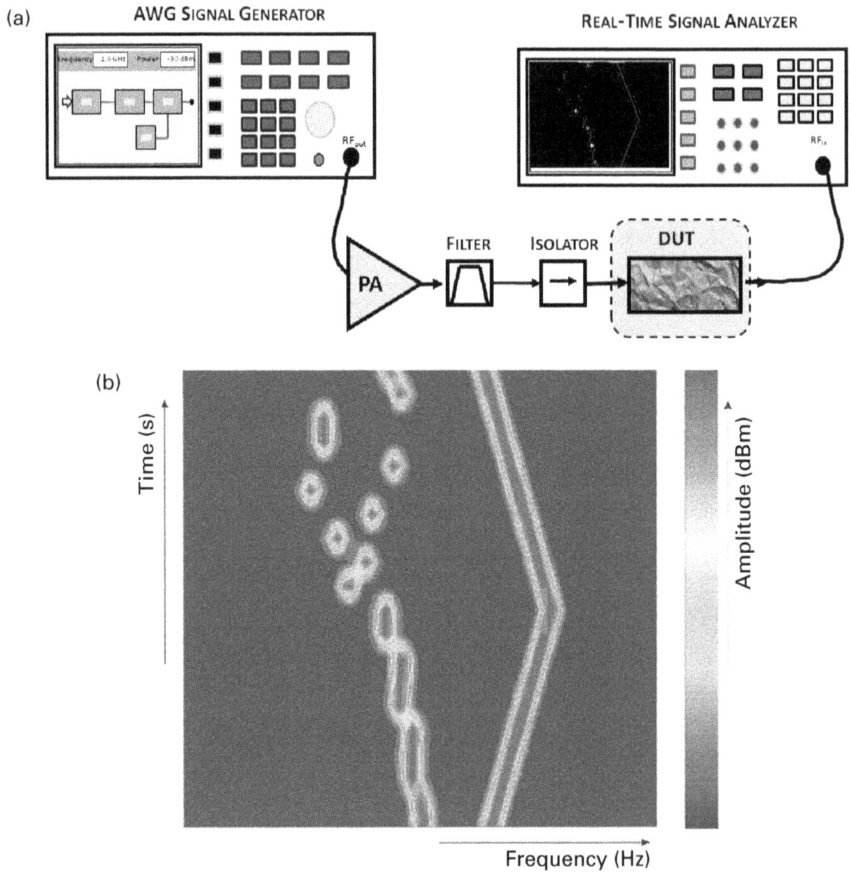


**Figure 4.30** Measurement of quantities over time: (a) an RTSA measurement bench and (b) a typical RTSA measurement.

DUT, one side is analog but the other is digital, which imposes that the measurements should be done in two different domains, one being digital and the another analog. Some vendors have already started thinking about it; for instance, note the equipment presented in [20, 21], which is an instrument similar to an oscilloscope, but one of the inputs is analog and the other one is digital, as in the logic analyzer presented in Section 2.9.

In these scenarios the information gathered is the same, meaning that all RF or baseband figures of merit could continue to be obtained, but now with a mixed-signal flavor. In the laboratory, users can rely on the instrumentation available on the market, or build their own using different instruments. Figure 4.31 presents a possible configuration.

The most important limitation in this scenario is the synchronization of all the instruments if a coherent measurement is to be done. More information on this type of measurement can be found in [20].

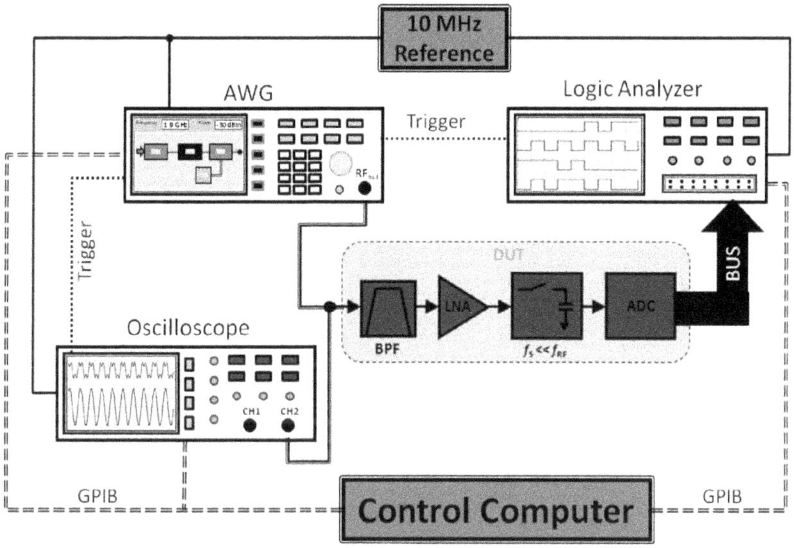


**Figure 4.31** A mixed-signal bench, with an AWG to generate the analog signal, an oscilloscope for capturing the analog signal and a logic analyzer for capturing the digital signals. All these measurements are then combined in a central computer for evaluation.

### 4.2.8 Temperature-dependent measurements

All the characterizations discussed in the above sections may be executed as functions of temperature in order to assess the circuit's performance under environmental conditions closer to those of practical operation. In the case of on-wafer measurements, a probe station whose chuck can be varied in temperature can be used. Special care should be taken about the calibration because probes and the standards on the calibration substrate change with temperature as well. In the case of packaged circuits, the primary focus of this book, the circuit is put in an oven that can be cooled and/or heated. An example is shown in Fig. 4.32. A GaN oscillator is put in an oven and its oscillation frequency is monitored by a spectrum analyzer while the temperature inside the oven is varied. The result is shown in Fig. 4.33.

## 4.3 Test benches for behavioral modeling

### 4.3.1 Introduction

In order to achieve wireless-system design, several circuit blocks can be cascaded. To reduce simulation time, the typical approach is to represent the various blocks by a behavioral model description instead of simulating the full circuit-level model which may consist of a hierarchical structure grouping tens of component models. From the wide range of behavioral models [22], we describe in this section the measurement benches of three major approaches: Volterra-series modeling, state-space modeling, and

**Figure 4.32** Measurements using a thermal chamber.

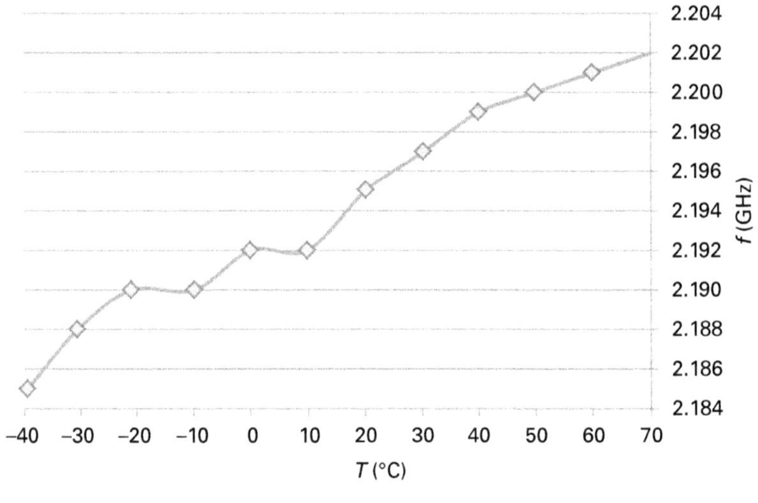


**Figure 4.33** The oscillation frequency of a GaN oscillator versus temperature.

the describing-functions approach. Whereas NVNA instruments can be adopted for any of these three methods, VSAs have traditionally been the instruments of preference for Volterra-series model extraction. The state-space modeling approach was originally developed on the basis of the NVNA, but also a VSA may be used if the modeling engineer is not interested in being able to predict the response around the harmonics. This is a valuable assumption, since one is often interested in the response at the carrier frequency only, assuming that the harmonic responses will be filtered out. The third approach, which is based on the describing-functions concept, has been developed directly in connection with the NVNA.

## 4.3.2 Volterra-series modeling

As seen in Section 1.4, nonlinear devices can be approximated using polynomial equations. This is especially true if the nonlinear device to be characterized and modeled is memoryless, meaning that it does not present dynamic effects. Nevertheless, sometimes nonlinear devices are also sensitive to nonlinear dynamics, which necessitates the use of more robust modeling strategies.

One possibility to model nonlinear dynamics is the so-called Volterra series [23], which somehow captures the behavior of the nonlinear device when the amplitude of the input signal is changed, but also when the carrier frequency (called short-term memory) or envelope dynamics (called long-term memory) are changed. A typical Volterra-series expansion can be expressed as

$$y(t) = \sum_{n=0}^{\infty} \int_{-\infty}^{\infty} \cdots \int_{-\infty}^{\infty} h_n(\tau_1, \ldots, \tau_n) x_{in}(t - \tau_1) \ldots x_{in}(t - \tau_n) d\tau_1 \ldots d\tau_n \quad (4.6)$$

where $h_n(\ldots)$ is called the $n$th-order Volterra kernel, and represents the behavior of an $n$th-order polynomial dynamic effect.

A Volterra series is thus a combination of linear convolution and a nonlinear power series that provides a general way to model a nonlinear system with memory. In that sense, it can be employed to describe the relationship between the input and output of a DUT presenting nonlinearities and memory effects.

For instance, in a third-degree polynomial, the Volterra kernel will be $h_3(\tau_1, \tau_2, \tau_3)$, and will thus represent the generation of the third-order nonlinear behavior that is dependent on a three-dimensional delay state vector.

If the Fourier transform of the Volterra kernel is taken, then the Volterra kernels represent thc nonlinear frequency response of the DUT nonlinearity. In this case they are called the nonlinear transfer functions. For instance, for a third-order nonlinearity and considering $N$ harmonics, the output can be represented as

$$\begin{aligned} y(t) = {} & \frac{1}{2} \sum_{n=-N}^{N} X(\omega_n) H_1(\omega_n) e^{j(\omega_n)t} \\ & + \frac{1}{4} \sum_{n_1=-N}^{N} \sum_{n_2=-N}^{N} X(\omega_{n_1}) X(\omega_{n_2}) H_2(\omega_{n_1}, \omega_{n_2}) e^{j(\omega_{n_1}+\omega_{n_2})t} \end{aligned}$$

$$+\frac{1}{8}\sum_{n_1=-N}^{N}\sum_{n_2=-N}^{N}\sum_{n_3=-N}^{N} X(\omega_{n_1})X(\omega_{n_2})X(\omega_{n_3})H_3(\omega_{n_1},\omega_{n_2},\omega_{n_3})$$
$$\times e^{j(\omega_{n_1}+\omega_{n_2}+\omega_{n_3})t} \tag{4.7}$$

The main problem with this formulation is how to obtain each $h_n(\ldots)$ for a specific nonlinear device. Moreover, the usual procedure described in the literature [24] is quite complex when one tries to extract all of the kernels at the same time, since this leads to an exponential increase in the number of coefficients with the degree of nonlinearity and memory length considered.

Moreover, it is well known [5] that sometimes the overall system description can behave very differently, since the even-order coefficients can generate signals at high frequencies (in the second-harmonic cluster) and at the baseband frequency near the DC cluster. This implies that the simultaneous extraction of all of the kernels is quite difficult since the formulation as presented in Eqs. (4.6) and (4.7) uses the same descriptor for the second harmonic as for the baseband (DC) response.

Thus a better way to extract and evaluate nonlinear Volterra kernels is by extracting each kernel while considering each nonlinear cluster as a low-pass equivalent signal, by which means each nonlinear mixing cluster is first selected and then converted individually to its complex envelope representation. The Volterra low-pass equivalent behavioral model will then be applied individually to each low-pass cluster [25]. This procedure is similar to what is being used in the envelope transient harmonic balance as described in [23]. Figure 4.34 presents this behavior.

Following this approach, each cluster will be extracted individually, and consequently the overall Volterra model will be the collection of each individual low-pass equivalent

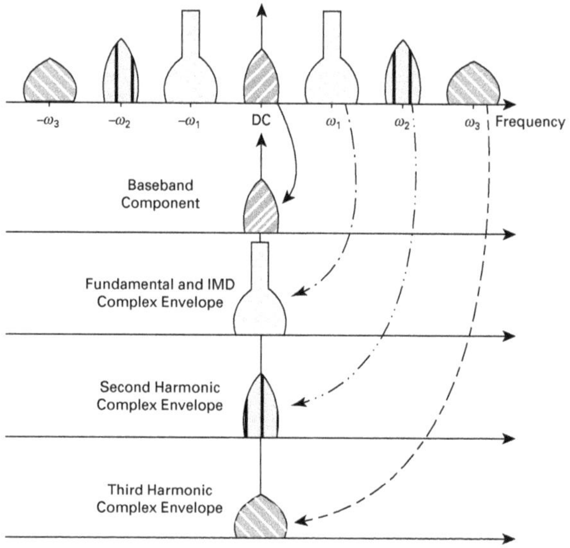


**Figure 4.34** Nonlinear clusters and bandpass equivalents, baseband, fundamental, and second and third harmonics. © IEEE.

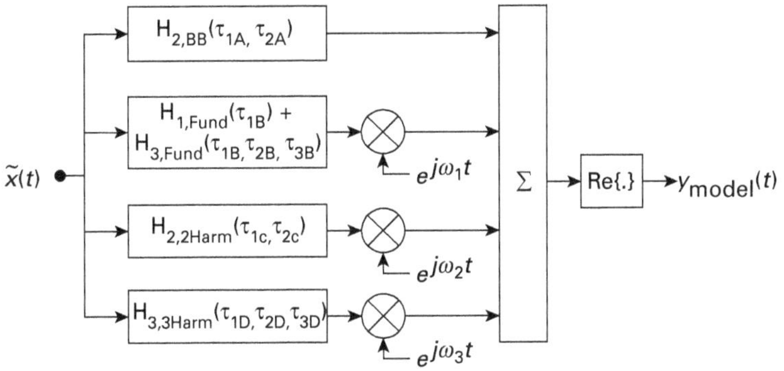


**Figure 4.35** The equivalent Volterra model, where each cluster can be identified. © IEEE.

model. It should be clear that we assume that the clusters do not overlap, which is the same as saying that this process can be applied to narrowband signals.

Figure 4.35 presents the implementation of the overall model, where the input signal $x(t)$ is first converted to its low-pass equivalent, which is called the complex waveform, $\tilde{x}(t)$, and then passed individually to each cluster. The output is then a collection of complex envelopes for each cluster $\tilde{y}_n(t)$, which should afterwards be up-converted to each frequency position and summed.

So for each cluster we will have the following. The NLTF for the baseband cluster is given by

$$\tilde{y}_{\mathrm{BB}}(k) = h_0 + \sum_{q_1=0}^{Q_{\mathrm{A1}}} \sum_{q_2=0}^{Q_{\mathrm{A2}}} \tilde{h}_{2,\mathrm{BB}}(q_1, q_2)\tilde{x}(k-q_1)\tilde{x}^*(k-q_2) \tag{4.8}$$

The NLTF for the fundamental cluster is

$$\begin{aligned} \tilde{y}_{\mathrm{fund}}(k) &= \sum_{q_1=0}^{Q_{\mathrm{B1}}} \tilde{h}_{1,\mathrm{fund}}(q_1)\tilde{x}(k-q_1) \\ &\quad + \sum_{q_1=0}^{Q_{\mathrm{B1}}} \sum_{q_2=0}^{Q_{\mathrm{B2}}} \sum_{q_3=0}^{Q_{\mathrm{B3}}} \tilde{h}_{3,\mathrm{fund}}(q_1, q_2, q_3)\tilde{x}(k-q_1)\tilde{x}(k-q_2)\tilde{x}^*(k-q_3) \end{aligned} \tag{4.9}$$

The NLTF for the second-harmonic cluster is

$$\tilde{y}_{\mathrm{2harm}}(k) = \sum_{q_1=0}^{Q_{\mathrm{C1}}} \sum_{q_2=0}^{Q_{\mathrm{C2}}} \tilde{h}_{2,\mathrm{2harm}}(q_1, q_2)\tilde{x}(k-q_1)\tilde{x}(k-q_2) \tag{4.10}$$

The NLTF for the third-harmonic cluster is

$$\tilde{y}_{\mathrm{3harm}}(k) = \sum_{q_1=0}^{Q_{\mathrm{D1}}} \sum_{q_2=0}^{Q_{\mathrm{D2}}} \sum_{q_3=0}^{Q_{\mathrm{D3}}} \tilde{h}_{3,\mathrm{3harm}}(q_1, q_2, q_3)\tilde{x}(k-q_1)\tilde{x}(k-q_2)\tilde{x}(k-q_3) \tag{4.11}$$

where $h_0$ is the DC value of the output, and $h_{2,\mathrm{BB}}$ and $h_{2,\mathrm{2harm}}$ are the second-order Volterra kernels for the baseband and second-harmonic responses, respectively. The tilde (˜) represents a complex signal or value, and the symbol $*$ denotes the complex conjugate. Different memory lengths are also visible for each cluster, $Q_q$, representing different dynamic effects.

Other clusters can be added for higher-degree Volterra kernels. It should be stressed here also that for each cluster all the nonlinear contributions should be added, as can be seen for instance in the fundamental cluster description in Eq. (4.9), where the linear kernel and the third-order nonlinear kernel are summed.

The signals corresponding to each of the clusters are in their complex-envelope format, and should then be up-converted to each cluster center frequency and added together, as in Fig. 4.35.

### 4.3.2.1 The parameter-extraction procedure

In order to extract the coefficients for the model described, a nonlinear DUT will be used for demonstration purposes. In this case the input should be as equal as possible to the signal we expect to have in the real environment where the DUT will be used. For gathering the data, in this case the time-domain samples $\tilde{x}(t)$, a wideband oscilloscope can be used to capture the overall signal, but in that case a reduced dynamic range will be obtained. Alternatively, a VSA can be used as the best solution, since the complex-envelope waveform description $\tilde{x}(t)$ is readily available in a VSA, and similarly all the subsequent output envelopes near each frequency cluster, $\tilde{y}_{\text{BB}}$, $\tilde{y}_{\text{fund}}$, $\tilde{y}_{\text{2harm}}$, and $\tilde{y}_{\text{3harm}}$, can be gathered sequentially.

Figure 4.36 presents the measurement bench used for this VSA-based extraction procedure.

The parameter-extraction approach will then be as follows.

1. Obtain the complex envelope for each cluster of the rearranged output signals; this can be done by selecting the correct central frequency for each cluster in the VSA.

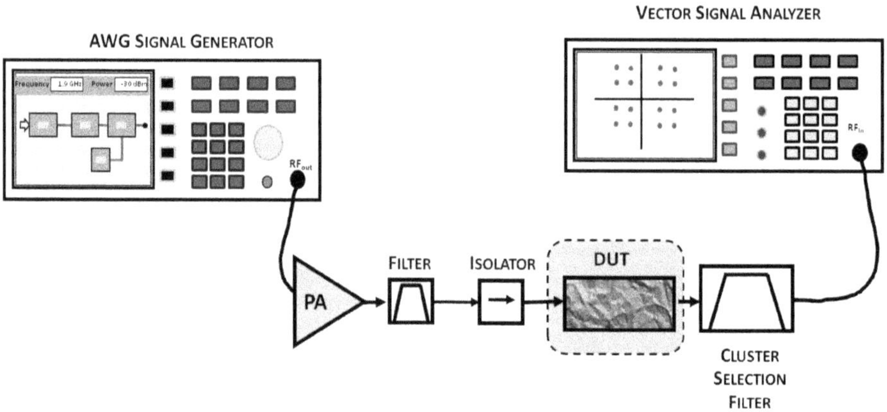


**Figure 4.36** A parameter-extraction measurement bench, where an extra filter is included to capture each cluster in the output spectrum.

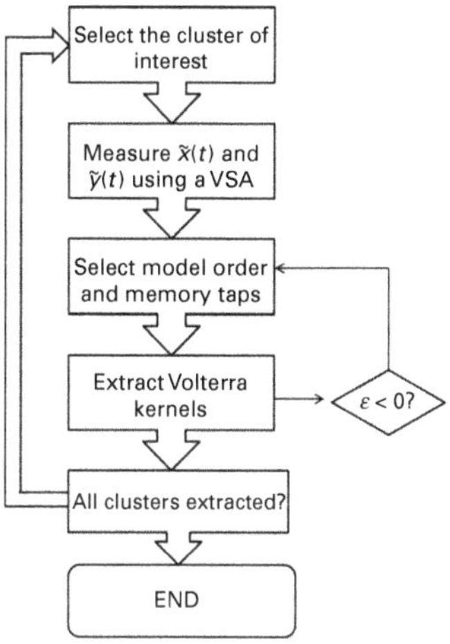


**Figure 4.37** A flowchart of the kernel-extraction procedure; $\varepsilon$ is the error goal for each cluster approximation.

2. Apply the low-pass equivalent Volterra-series model, Eqs. (4.8)–(4.11), to these new output signals using also the measured input complex envelope and obtain the desired low-pass complex Volterra kernels. (This is readily done if a VSA is used.)
3. Up-convert each output complex signal to the corresponding cluster center frequency and finally assess the model performance.

Figure 4.37 presents the followed approach.

It should be clear here that the input signal and the output signal should be synchronized. This can be achieved by, for instance, embedding a trigger into the signal as in [25], or by using external triggers. If the solution is to use an embedded trigger, then the signal to be fed to the DUT should be a signal in which the first samples contain the trigger itself, as illustrated in Fig. 4.38. These initial samples should be used for the triggering strategy and then deleted prior to the procedure for extraction of the coefficients.

As mentioned previously, we are able to decide on the nonlinear order and number of memory taps that are more convenient for each cluster. In this way the overall number of parameters required in order to match the output signals can be reduced by using this separate processing. In order to reduce even more the impact of the measurement noise, several independent measurements can be taken from the output signals and then averaged, thereby diminishing the significance of the noise level [25].

Considering now that the input signal is sufficiently rich, i.e., one that presents a high variability, and that the output of the Volterra-series model is linear with respect to its parameters, several low-pass complex Volterra kernels can be determined using a least-squares technique, as expressed by

$$H = (X^{\mathrm{T}}X)^{-1}X^{\mathrm{T}}Y \tag{4.12}$$

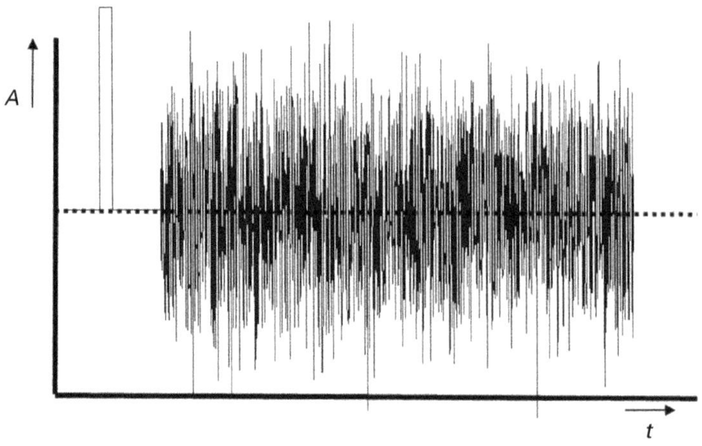


**Figure 4.38** An embedded trigger.

where $X$ and $Y$ are the input complex signal matrix and the output signal vector, respectively, and $H$ is the vector of complex kernels that we are looking for. So, this least-squares extraction has to be performed for each of the clusters selected. In order to understand this mechanism, consider first a memoryless device. In this case the output can be calculated as a polynomial sum,

$$y(k) = \sum_{p=0}^{P} a_p x(k)^p \tag{4.13}$$

and the extraction of each $a_p$ can be done using the simple equation

$$y(k) = \begin{bmatrix} y(0) \\ \cdots \\ y(n) \\ \cdots \\ y(N) \end{bmatrix} = \begin{bmatrix} 1 & x(0) & \ldots x(0)^p & \ldots & x(0)^P & \vdots \\ \vdots & \vdots & & \vdots & & \vdots \\ 1 & x(n) & \ldots x(n)^p & \ldots & x(n)^P & \\ \vdots & \vdots & & \vdots & & \vdots \\ 1 & x(N) & \ldots x(N)^p & \ldots & x(N)^P & \end{bmatrix} \begin{bmatrix} a_0 \\ \cdots \\ a_p \\ \cdots \\ a_P \end{bmatrix} \tag{4.14}$$

with

$$A = (X^{\mathrm{T}} X)^{-1} X^{\mathrm{T}} Y \tag{4.15}$$

The coefficients $a_p$ can then be calculated using Eq. (4.12).

If the polynomial is not memoryless but presents memory effects, then some degree of memory should be included, and the overall impulse response calculated for each cluster. Considering as an example that the complex parameters for the baseband cluster's second-order nonlinearity are to be found considering a memory length of $Q$ taps, the input signal matrix ($X$) should be composed by taking

$$\mathbf{X} = \begin{bmatrix} 1 & \tilde{x}(0)\tilde{x}^*(0) & \tilde{x}(0)\tilde{x}^*(-q) & \dots & \tilde{x}(-Q)\tilde{x}^*(-Q) \\ \vdots & \vdots & \vdots & & \vdots \\ 1 & \tilde{x}(n)\tilde{x}^*(n) & \tilde{x}(n)\tilde{x}^*(n-q) & \dots & \tilde{x}(n-Q)\tilde{x}^*(n-Q) \\ \vdots & \vdots & \vdots & & \vdots \\ 1 & \tilde{x}(N)\tilde{x}^*(N) & \tilde{x}(N)\tilde{x}^*(N-q) & \dots & \tilde{x}(N-Q)\tilde{x}^*(N-Q) \end{bmatrix} \tag{4.16}$$

and $Y_{\mathrm{BB}}$, the complex output at baseband frequencies, is obtained as

$$\mathbf{Y}_{\mathrm{BB}} = \left[\tilde{y}_{\mathrm{BB}}(0) \ \dots \ \tilde{y}_{\mathrm{BB}}(n) \ \dots \ \tilde{y}_{\mathrm{BB}}(N)\right]^{\mathrm{T}} \tag{4.17}$$

where $Q$ represents the memory length and $N$ is the number of samples captured for the input complex-envelope signal and for the output signals. The entries in the first column of the matrix in Eq. (4.16) are all 1s, due to the DC component.

Thus $H$ can then be calculated using Eq. (4.12). This result has the advantage of notational simplicity and general applicability. $H$ is actually composed by the following Volterra operators:

$$\mathbf{H}_{\mathrm{BB}} = \left[\tilde{h}(0) \ \tilde{h}_{2,\mathrm{BB}}(0,0) \ \dots \ \tilde{h}_{2,\mathrm{BB}}(Q,Q)\right]^{\mathrm{T}} \tag{4.18}$$

This procedure should be implemented for the next clusters. When all coefficients are extracted, they should then be up-converted to the corresponding frequency cluster using

$$y(k) = \mathrm{Re}\{\tilde{y}_{\mathrm{BB}}(k) + \tilde{y}_{\mathrm{fund}}(k)e^{j\omega_1 t} + \tilde{y}_{\mathrm{2harm}}(k)e^{j2\omega_1 t} + \tilde{y}_{\mathrm{3harm}}(k)e^{j3\omega_1 t}\} \tag{4.19}$$

In this way a nonlinear black-box model can be extracted using a simple and efficient approach. It should also be mentioned that these models behave well for the signal used in the extraction, but can deviate from the real values for similar signals with higher amplitudes. If a model description for a large range of amplitudes, or powers, has to be obtained, then the model-extraction procedure should be repeated for each amplitude of interest. More information on the applicability of the Voterra series can be found in [24].

### 4.3.3 State-space modeling

State-space modeling is a widely spread modeling technique in engineering. It is a long-standing approach in mechanical and chemical engineering, as well as in low-frequency electronics. But it was not introduced into microwave engineering until the NVNA had been developed [26], because the NVNA made it possible that the responses of a non-linear microwave two-port (and later on three-port) system could be fully characterized, namely both in amplitude and in phase.

The general model formulation expressed in the time domain is

$$\begin{aligned} \dot{X}(t) &= f_{\mathrm{a}}(X(t), U(t)) \\ Y(t) &= f_{\mathrm{b}}(X(t), U(t)) \end{aligned} \tag{4.20}$$

with $U(t)$ the vector of inputs, $X(t)$ the vector of state variables, and $Y(t)$ the vector of outputs. The overdot denotes the time derivative. The functions $f_a(.)$ and $f_b(.)$ are analytical functions expressing the dependencies.

In applying the concept of state-space modeling to microwave devices, two approaches are adopted, depending on the actual application. Either the inputs and outputs can be set to the port voltages and currents, or they can be expressed in terms of the incident and scattered waves, as in the following expression for a two-port DUT:

$$\begin{aligned} b_1(t) &= f_1\left(a_1(t), \dot{a}_1(t), \ddot{a}_1(t), \ldots, a_2(t), \dot{a}_2(t), \ddot{a}_2(t), \ldots, \dot{b}_1(t), \ldots, \dot{b}_2t, \ldots\right) \\ b_2(t) &= f_2\left(a_1(t), \dot{a}_1(t), \ddot{a}_1(t), \ldots, a_2(t), \dot{a}_2(t), \ddot{a}_2(t), \ldots, \dot{b}_1(t), \ldots, \dot{b}_2t, \ldots\right) \end{aligned} \tag{4.21}$$

Note that Eq. (4.21) is applicable only to devices that exhibit no long-term memory effects. The reader is referred to [27] for the extension of the model formulation to include long-term memory effects.

The modeling sequence is visualized in Fig. 4.40. First some initial measurements have to be performed such that the order of dynamics, or in other words the derivatives' order, can be determined. A common approach is the so-called false-nearest-neighbors method [26]. It is essential for the subsequent steps that this set of initial measurements covers the experimental conditions (frequency range, bias range, power range) for which the resulting model should be valid in the end. For example, if the order of dynamics is established on the basis of NVNA measurements at 500 MHz, whereas the aim is to have a model valid at 10 GHz, the 500-MHz data might not reveal all of the dynamics in the DUT's behavior at 10 GHz. Once the order of the dynamics has been established, a full set of measurements has to be performed so as to cover the multi-dimensional state space as well as possible. This is followed by the actual model construction, which is involved in determining the functions $f_1$ and $f_2$ in Eq. (4.21). These are analytical expressions, e.g., artificial neural networks, whose parameters are determined by optimization against the collected measurement data. Next, the model obtained is validated, and, if the accuracy is not adequate, the sequence is repeated, possibly starting by collecting additional measurements, and revisiting the model parameters' optimization step. Since experiment design is a crucial step, some more details are provided in the following.

The aim of experiment design is to cover the state space with measurement data, such that the functions $f_1(.)$ and $f_2(.)$ can be determined. A dense and complete coverage is a necessity in order to avoid the need for interpolation and extrapolation when using the resulting model in system-level design or other applications.

Using single-tone measurements would quickly lead to a high amount of measurements at various experimental conditions (range of biases, carrier frequencies, and power levels) in order to obtain the required dense coverage of the state space. Instead, it is more measurement-efficient to use multi-tone excitations for which the tone spacing is small compared with the RF carrier frequency [28]. The best excitation would be a random noise source [29], but this is practically difficult to achieve in high-frequency measurement, especially considering that the operation of an NVNA requires periodic signals. So, in practice, a multi-sine excitation or a set

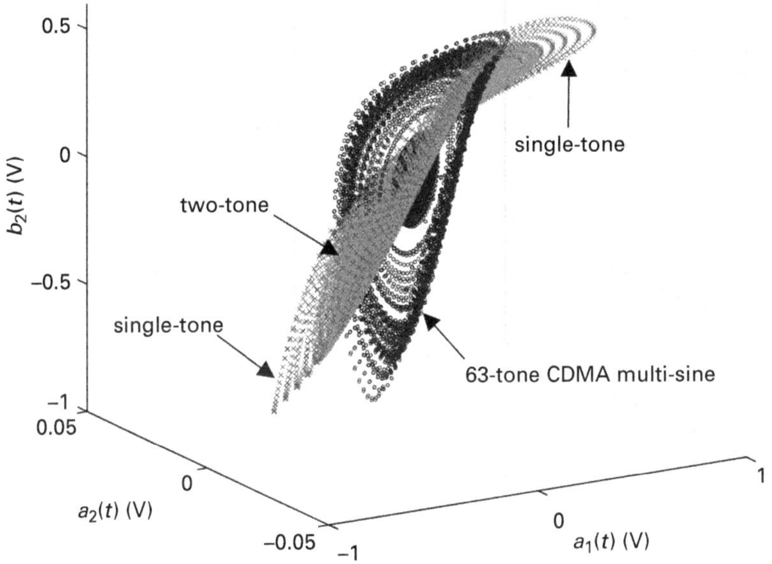


**Figure 4.39** The $(a_1(t), a_2(t), b_2(t))$ coverage for a power-swept single-tone measurement, a two-tone measurement, and a 63-tone CDMA-like multi-sine measurement on an off-the-shelf amplifier. © IEEE.

of multi-sine excitations is used to cover the state space. As we deviate from the ideal solution, namely a random noise source, the choice of multi-sine is correlated with the DUT and its application range [30, 31]. This is illustrated by the following example.

An off-the-shelf amplifier has been modeled on the basis of two data sets. The first model is based on power-swept single-tone measurements, and is called "Model 1," whereas the second model is based on a measurement involving a 63-tone multi-sine excitation, and is called "Model 63." Figure 4.39 shows the corresponding coverage in the $(a_1(t), a_2(t), b_2(t))$ space. Note that this is a subprojection of the actual state space, since this three-dimensional representation does not yet show any higher-order dynamics. Figure 4.39 also includes the coverage corresponding to a two-tone excitation.

In the first instance, one may conclude from the analysis above that a more complex multi-sine excitation always results in better model accuracy. The counterexample can be illustrated with this example. Intuitively, one may expect that Model 63, which is based on a multi-sine, would be able to predict the two-tone measurement results, which are also of multi-sine nature, the best. But the contrary is correct. In this particular example, Model 1 actually performs considerably better [31]. The explanation can be deduced from Fig. 4.39. It can be observed that the coverage of the set of single-tone measurements is very similar to the coverage of the two-tone measurement. In the case of Model 63, extrapolation is required in order to predict the response for the

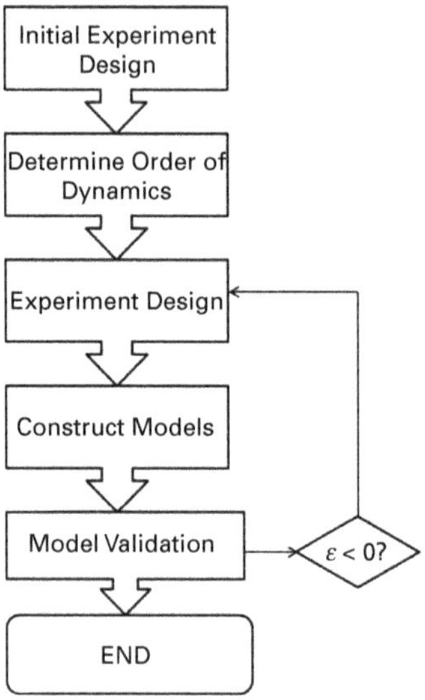


**Figure 4.40** The sequence of steps employed to construct a state-space model.

two-tone excitation, and extrapolation usually results in larger inaccuracies. This example demonstrates that the coverage of the state space is crucial for constructing good state-space behavioral models.

An illustration of the accuracy of state-space behavioral models can be seen in Fig. 4.19.

In terms of applicability, state-space models are especially suitable for those modeling tasks where the DUT exhibits strongly nonlinear behavior. The reason is that Volterra-series models are usually truncated to nonlinear order three, due to the fact that the number of model parameters grows quickly once the nonlinearity order exceeds three. So the most common use of Volterra-series models is for weakly nonlinear DUTs. Similarly, state-space models are not bound by the limiting conditions under which the $S$-function modeling concept (see Section 4.3.4) is valid. On the other hand, the drawback of state-space modeling is that the optimal experiment design is strongly DUT-dependent and therefore requires user insight and experience. Also, model construction is based on optimization, and this is prone to the usual pitfalls, such as local minima and overtraining.

### 4.3.4 Beyond *S*-parameters

A recent modeling technique is an extension to the concept of $S$-parameters. It is based on the describing-functions concept. Over the years, slightly different formulations have been proposed in the literature, such as $X^{\text{TM}}$-parameters and $S$-functions. In the following, the expressions used correspond to the $S$-function description.

The principle of this approach is that the response of the DUT is linearized around a large-signal operating condition. It is illustrated in Fig. 4.41. The result at port 2 of

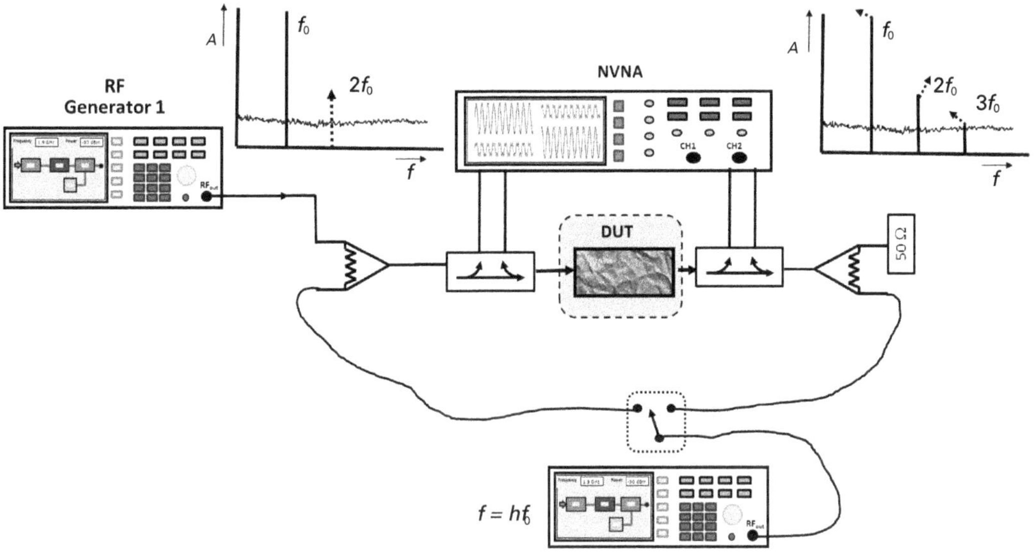


**Figure 4.41** A schematic representation explaining the beyond-$S$-parameters concept.

applying a large-signal single-tone excitation at port 1 at frequency $f_0$ is indicated by the spectrum plot on the right-hand side. This state determines the large-signal operating condition. Note that the 50-Ω port at port 2 can be replaced by a tuner, so as to realize a non-50-Ω large-signal operating condition. When a small probing signal is applied at a harmonic frequency (or at the fundamental frequency at port 2), the result is such that there is a linear response at each of the other spectral components. Since the device response is dependent on the phase relationship between the small probing signal and the large-signal excitation, the corresponding change in the device response cannot be described by a simple linear relationship with the incident phasors, and therefore additional terms need to be included in the model. This corresponds to the introduction of the conjugate terms "$c$" in the formulation [32].

In the $S$-function modeling approach, the scattered traveling voltage waves $b_1$ and $b_2$ are expressed as functions of the incident traveling voltage waves $a_1$ and $a_2$, as follows:

$$b'_{ph} = Sf_{ph11}|a_{11}| + \sum_{ij\neq 11}\left(Sf_{phij}a'_{ij} + Sfc_{phij}\left(a'_{ij}\right)^{*}\right) \tag{4.22}$$

where $p$ and $h$ denote the output port and harmonic indices, respectively, and $i$ and $j$ stand for the input port and harmonic indices, respectively, at which the probing signal is applied. The superscript * stands for the conjugate operator, and the primes denote a phase shift relative to the phase of $a_1$ at the carrier frequency:

$$x'_{ph} = x' \exp(-jh\phi(a_{11})) \tag{4.23}$$

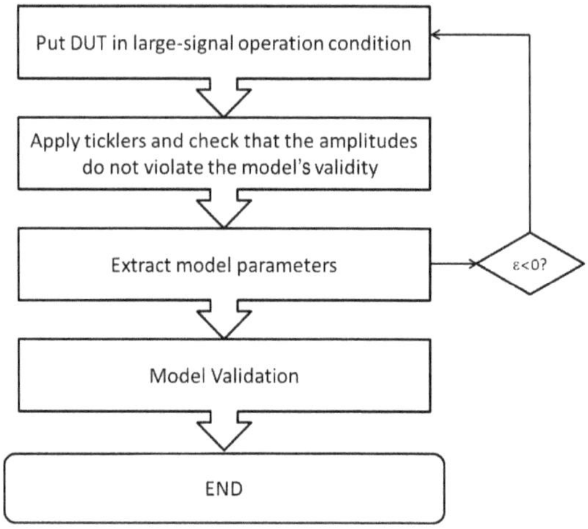


**Figure 4.42** The sequence of steps employed to construct the beyond-$S$-parameters model.

The complex coefficients $Sf_{phij}$ and $Sfc_{phij}$ are called $S$-functions [33]. Note that, for simplicity, the DC part of the device's response has been neglected in Eq. (4.22).

The modeling sequence is depicted in Fig. 4.42.

The measurement procedure is started by putting the device under a large-signal operating condition of interest, i.e., fixing the bias condition, carrier frequency, and input power level. It may be that the output load is not kept at the standard 50 Ω, but that the model is to be constructed for another load condition. In other words, load-pull is applied simultaneously. It has been explained in Section 4.2.5.2 that fundamental load-pull corresponds to $a_2$ not being zero at the fundamental frequency. In such a case, the non-zero $a_2$ is part of the large-signal operation condition as well. This concept can be extended to harmonic load-pull as well.

The next step consists of applying the so-called ticklers. This means that a small probing signal is sequentially injected at each of the harmonic frequencies at port 1 and at port 2. Also, if there is no load-pull, a measurement by which a small probing signal at the fundamental frequency $f_0$ is injected at port 2 has to be added.

These measurements can be easily automated, so it is possible that the user may forget about the underlying assumptions. The model approach under consideration relies on the fact that the probing excitations result in a linear response at each of the spectral components.

A straightforward test is to increase the amplitude of the small probing signal and observe the resulting change in the corresponding scattered traveling voltage wave phasors. To facilitate the analysis, the small probing signals may be applied at a frequency that is slightly offset, by $\Delta f$ [33], as illustrated in Fig. 4.43. As a result, also the small-signal and conjugate small-signal contributions are separated in frequency, and, moreover, do not coincide with the spectral components corresponding to the large-signal operating condition. For the linearization assumption to hold, it is required that

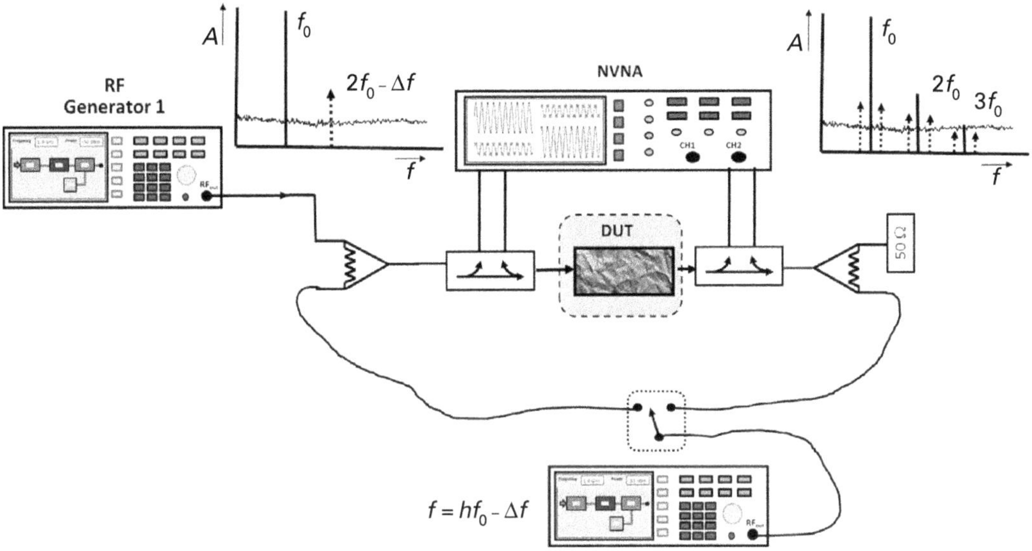


**Figure 4.43** A tickler is applied at a frequency offset.

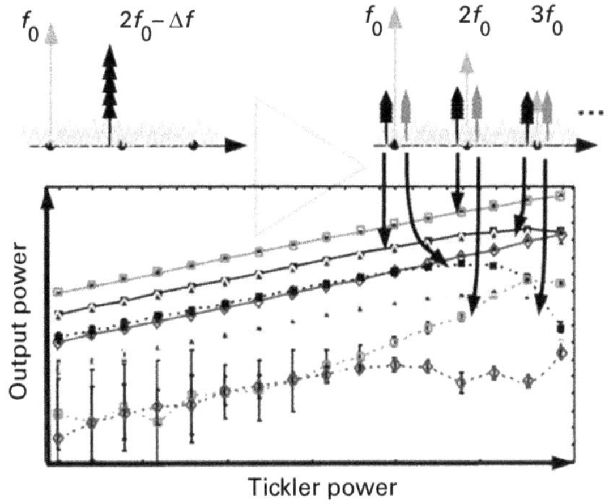


**Figure 4.44** Checking whether the newly created spectral components behave linearly as functions of tickler amplitude. The DUT is a packaged transistor [34]. © IEEE.

the magnitude of the response at the offset frequencies $hf_0 - \Delta f$ and their conjugates $hf_0 + \Delta f$ change linearly, while the magnitude of the response on the fundamental frequency grid $hf_0$ should remain unchanged [34]. This is illustrated for measurements on a packaged transistor in Figs. 4.44 and 4.45, respectively. Also, it is not a solution to make the probing signal very small, since the other limit in terms of amplitude value is the noise floor. Since the conjugate small-signal contributions in the example in Fig. 4.44 are small with small tickler amplitude, it is not straightforward to conclude

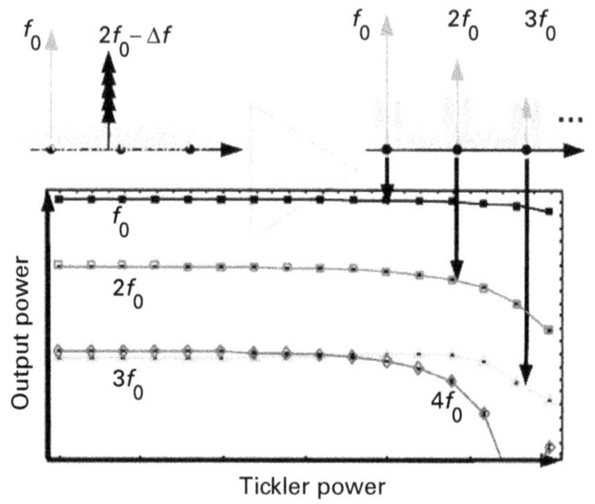


**Figure 4.45** Checking whether the large-signal operating condition remains constant with increasing tickler amplitude. The DUT is a packaged transistor [34]. © IEEE.

whether any deviations from the linear behavior are due to measurement uncertainty or an indication of violating the assumption of linearity. This can be resolved by repeating the measurements and calculating the variance for the measured spectral components, as represented by the vertical bars in Fig. 4.44.

Not only should the magnitude of the spectral components be checked, but also it has to be verified that the phase relationship between the response at $hf_0$ and at $hf_0 \pm \Delta f$ and the respective large- and small-signal inputs remains constant. Note that, in the case of the response at the conjugate frequencies $hf_0 + \Delta f$, the phase differences are calculated with respect to the inverted phase of the probe tone. Finally, the probing signal level must not cause any response at the frequencies corresponding to higher-order intermodulation products of $hf_0$ and $hf_0 - \Delta f$ components. In general one seeks the maximum amplitude of the probing signal at which all these assumptions are valid. Note that this should be checked for each of the probing signals, due to the frequency-dependent behavior of the DUT.

Once it has been checked that the experimental conditions do not lead to a violation of the method's validity, the actual data for model construction can be collected. For each of the probing-signal excitations, the corresponding responses at all fundamental and harmonic spectral components are measured. To be able to distinguish between $Sf_{phij}$ and $Sfc_{phij}$, at least two measurements, namely two different phase values for the probing signal, are required for each of the settings. Once all of the measurements have been collected, the $S$-function parameters can be determined by solving a linear set of equations, starting from Eq. (4.22). Owing to measurement imperfections, such as noise, the presence of harmonics generated by the sources, nonlinear interaction between source and DUT, and imperfect terminations, an optimization procedure is usually required. An alternative way to obtain the $S$-function parameters is by applying the ticklers again at a frequency that is slightly offset [33]. The advantage of this approach is that it is no longer the case that all of the measurements have to be collected before the $S$-function parameters can be determined by solving the set of linear equations. Instead, the $S$-function parameters can be directly determined from a partial set of measurements, i.e., up to the harmonic of interest, making the overall measurement time shorter.

Finally, it should be noted that the model parameters, namely the $S$-functions, are linked to a particular large-signal operating condition, and therefore the whole process of model construction has to be repeated for each large-signal operation condition of interest. In practice, this measurement-and-extraction sequence is provided as an option in the NVNA instrument by the manufacturer, so it is largely a "push-button" solution for the user.

## Problems

**4.1** Discuss the main problems in the measurements when using a two-tone test with an AWG.

**4.2** Consider a system that has an input described by $x(t) = [11,\ 21,\ 31,\ 520,\ 10, 2]$, and an output described by $y(t) = [22,\ 642,\ 993,\ 7550,\ 110,\ 24]$. Calculate the linear and second-order Volterra kernel, considering that the system is memoryless.

**4.3** If a system presents memory, describe the best strategy to account for that behavior in a two-tone measurement.

**4.4** Describe the best strategy to evaluate the SNR degradation using a VSA.

**4.5** What are the differences between the tuning of the CCPR measurement bench in small-signal operation and that in large-signal operation?

**4.6** What are the main differences between measuring spectra using a spectrum analyzer and doing so using an oscilloscope?

**4.7** Describe the main differences between the VNA and the NVNA measurement bench.

**4.8** Explain the different approaches for measuring large-signal $S$-parameters when using the tickler method.

## References

[1] N. B. Carvalho, J. C. Pedro, and J. P. Martins, "A corrected microwave multisine waveform generator," *IEEE Trans. Microwave Theory Tech.*, vol. 54, no. 6, part 2, pp. 2659–2664, Jun. 2006.

[2] N. Suematsu, Y. Lyama, and O. Ishida, "Transfer characteristic of IM relative phase for a GaAs FET amplifer," *IEEE Trans. Microwave Theory Tech.*, vol. 45, no. 12, pp. 2509–2514, Dec. 1997.

[3] Y. Yang, J. Yi, J. Nam, B. Kim, and M. Park, "Measurement of two-tone transfer characteristics of high-power amplifers," *IEEE Trans. Microwave Theory Tech.*, vol. 49, no. 3, pp. 568–571, Mar. 2001.

[4] J. P. Martins and N. B. Carvalho, "Multi-tone phase and amplitude measurement for nonlinear device characterization," *IEEE Trans. Microwave Theory Tech.*, vol. 53, no. 6, pp. 1982–1989, Jun. 2005.

[5] N. B. Carvalho and J. C. Pedro, "A comprehensive explanation of distortion side band asymmetries," *IEEE Trans. Microwave Theory Tech.*, vol. 50, no. 9, pp. 2090–2101, Sep. 2002.

[6] R. Liu, D. Schreurs, W. De Raedt, F. Vanaverbeke, and R. Mertens, "RF-MEMS based tri-band GaN power amplifier," *Electronics Lett.*, vol. 47, no. 13, pp. 762–763, Jun. 2011.

[7] M. A. Yarlequé Medina, D. Schreurs, and B. Nauwelaers, "RF class-E power amplifier design based on a load line–equivalent capacitance method," *IEEE Microwave Wireless Components Lett.*, vol. 18, no. 3, pp. 206–208, Mar. 2008.

[8] J. Dunsmore, "Novel method for vector mixer characterization and mixer test system vector error correction," in *IEEE MTT-S International Microwave Symposium*, Jun. 2002, pp. 1833–1836.

[9] J. Dunsmore, "A new calibration method for mixer delay measurements that requires no calibration mixer," in *European Microwave Conference (EuMC)*, Oct. 2011, pp. 480–483.

[10] G. Pailloncy, G. Avolio, M. Myśliński *et al.*, "Large-signal network analysis including the baseband," *IEEE Microwave Mag.*, vol. 12, no. 2, pp. 77–86, Apr. 2011.

[11] G. Crupi, G. Avolio, D. Schreurs *et al.*, "Vector two-tone measurements for validation of nonlinear microwave FINFET model," *Microelectronic Eng.*, vol. 87, no. 10, pp. 2008–2013, Oct. 2010.

[12] N. B. Carvalho, K. A. Remley, D. Schreurs, and K. G. Gard, "Multisine signals for wireless system test and design," *IEEE Microwave Mag.*, vol. 9, no. 3, pp. 122–138, Jun. 2008.

[13] M. Myśliński, D. Schreurs, and B. Nauwelaers, "Large-signal time-domain behavioral model of a packaged high efficient wireless circuit design," in *European Microwave Conference (EuMC)*, Oct. 2004, pp. 565–568.

[14] F. F. Vanaverbeke, W. W. De Raedt, D. D. Schreurs, and M. M. Vanden Bossche, "Real-time non-linear de-embedding," in *Automatic RF Techniques Group Conference*, Jun. 2011, pp. 1–6.

[15] P. J. Tasker, "Practical waveform engineering," *IEEE Microwave Magazine*, vol. 10, no. 7, pp. 65–76, Dec. 2009.

[16] A. Raffo, G. Avolio, D. Schreurs *et al.*, "On the evaluation of the high-frequency load line in active devices," *Int. J. Microwave Wireless Tech.*, vol. 3, no. 1, pp. 19–24, Feb. 2011.

[17] J. Pedro and N. de Carvalho, "Characterizing nonlinear RF circuits for their in-band signal distortion," *IEEE Trans. Instrumentation Measurement*, vol. 51, no. 3, pp. 420–426, Jun 2002.

[18] H. Ku, W. Woo, and J. Kenney, "Carrier-to-interference ratio prediction of nonlinear RF devices," *Microwave J.*, vol. 44, no. 2, pp. 154–164, 2001.
[19] C. Ghosh, S. Roy, and D. Cavalcanti, "Coexistence challenges for heterogeneous cognitive wireless networks in TV white spaces," *IEEE Trans. Wireless Communications*, vol. 18, no. 4, pp. 22–31, Aug. 2011.
[20] P. Cruz, N. Carvalho, and K. Remley, "Designing and testing software-defined radios," *IEEE Microwave Mag.*, vol. 11, no. 4, pp. 83–94, Jun. 2010.
[21] P. Cruz, N. Carvalho, K. Remley, and K. Gard, "Mixed analog–digital instrumentation for software-defined-radio characterization," in *2008 IEEE MTT-S International Microwave Symposium Digest* Jun. 2008, pp. 253–256.
[22] D. Schreurs, M. O'Droma, A. Goacher, and M. Gadringer, *Behavioural Modelling Techniques for RF Power Amplifiers*. Cambridge: Cambridge University Press, 2008.
[23] J. C. Pedro and N. B. Carvalho, *Intermodulation Distortion in Microwave and Wireless Circuits*. New York: Artech House, 2003.
[24] J. Pedro and S. Maas, "A comparative overview of microwave and wireless power-amplifier behavioral modeling approaches," *IEEE Trans. Microwave Theory Techniques*, vol. 53, no. 4, pp. 1150–1163, Apr. 2005.
[25] P. Cruz and N. Carvalho, "Wideband behavioral model for nonlinear operation of bandpass sampling receivers," *IEEE Trans. Microwave Theory Tech.*, vol. 59, no. 4, pp. 1006–1015, Apr. 2011.
[26] D. Schreurs, J. Wood, N. Tufillaro, L. Barford, and D. Root, "Construction of behavioural models for microwave devices from time-domain large-signal measurements to speed-up high-level design simulations," *Int. J. RF Microwave Computer Aided Eng. (RFMICAE)*, vol. 13, no. 1, pp. 54–61, Jan. 2003.
[27] D. Schreurs, K. Remley, M. Myśliński, and R. Vandersmissen, "State-space modelling of slow-memory effects based on multisine vector measurements," in *Automatic RF Techniques Group Conference (ARFTG)*, Dec. 2003, pp. 81–87.
[28] D. Schreurs and K. Remley, "Use of multisine signals for efficient behavioural modelling of RF circuits with short-memory effects," in *Automatic RF Techniques Group Conference (ARFTG)*, Jun. 2003, pp. 65–72.
[29] R. Pintelon and J. Schoukens, *System Identification: A Frequency Domain Approach*. New York: IEEE Press, 2001.
[30] D. Schreurs, M. Myśliński, and K. Remley, "RF behavioural modelling from multisine measurements: Influence of excitation type," in *European Microwave Conference (EuMC)*, Oct. 2003, pp. 1011–1014.
[31] M. Myśliński, D. Schreurs, and B. Nauwelaers, "Impact of sampling domain and number of samples on the accuracy of large-signal multisine measurement-based behavioral model," *Int. J. RF Microwave Computer Aided Eng. (RFMICAE)*, vol. 20, no. 4, pp. 374–380, May 2010.
[32] J. Verspecht and D. Root, "Polyharmonic distortion modeling," *IEEE Microwave Magazine*, vol. 7, no. 3, pp. 44–57, Jun. 2006.

[33] M. Myśliński, F. Verbeyst, M. Vanden Bossche, and D. Schreurs, "$S$-functions extracted from narrow-band modulated large-signal network analyzer measurements," in *Automatic RF Techniques Group Conference (ARFTG)*, Dec. 2009, p. 8.

[34] M. Myśliński, F. Verbeyst, M. Vanden Bossche and D. Schreurs, "A method to select correct stimuli levels for $S$-functions behavioral model extraction," in *IEEE MTT-S International Microwave Symposium*, Jun. 2010, pp. 1170–1173.

# Index

For EU product safety concerns, contact us at Calle de José Abascal, 56–1°,
28003 Madrid, Spain or eugpsr@cambridge.org.

www.ingramcontent.com/pod-product-compliance
Ingram Content Group UK Ltd.
Pitfield, Milton Keynes, MK11 3LW, UK
UKHW050620211026
15995UKWH00005B/100